D1634035

CONTENTS

Photo Credits: Front cover *Lamprologus brichardi* by H.J. Richter. Back cover *Melanochromis johanni* (both sexes). Front endpaper *Lamprologus leleupi* by H.J. Richter. Back endpaper *Pseudotropheus lanisticola* by Ken Lucas, Steinhart Aquarium. Title page *Haplochromis moorii* by G.Marcuse.

To Jennifer

ISBN 0-87666-514-8

© 1979 by T.F.H. Publications, Inc.

Distributed in the U.S. by T.F.H. Publications, Inc., 211 West Sylvania Avenue, P.O. Box 427, Neptune, N.J. 07753; in England by T.F.H. (Gt. Britain) Ltd., 13 Nutley Lane, Reigate, Surrey; in Canada to the book store and library trade by Beaverbooks, 953 Dillingham Road, Pickering, Ontario L1W 1Z7; in Canada to the pet trade by Rolf C. Hagen Ltd., 3225 Sartelon Street, Montreal 382, Quebec; in Southeast Asia by Y.W. Ong, 9 Lorong 36 Geylang, Singapore 14; in Australia and the South Pacific by Pet Imports Pty. Ltd., P.O. Box 149, Brookvale 2100, N.S.W., Australia; in South Africa by Valiant Publishers (Pty.) Ltd., P.O. Box 78236, Sandton City, 2146, South Africa; Published by T.F.H. Publications, Inc., Ltd., The British Crown Colony of Hong Kong.

RIFT LAKE CICHLIDS

BY GLEN S. AXELROD

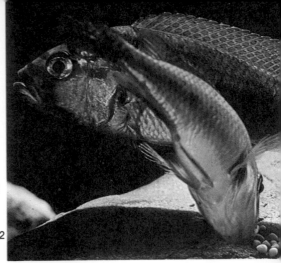

A spawning sequence
of *Haplochromis burtoni*
shows (1) buccal cavity,
(2) fertilization, (3-4)
female collecting eggs,
(5) male display, (6)
female discharging
developed young.

5

6

Preface

This book is intended as an introduction to Rift Lake cichlids and is not in any way a comprehensive or scientific work. Rift Lake cichlids have become popular and wide-spread in the aquarium hobby only within the past ten years, and there has been a growing need for introductory literature on this subject. As both a scientist and a hobbyist, I have attempted to blend aquarist information on Rift Lake cichlid care with a review of their origin and evolution. The information in this book comes from a review of the literature and experience gained in both the laboratory and field while working on Rift Lake cichlids in Africa for three years.

This book contains information on how to set up and maintain your aquarium and how to choose, feed and breed your fish. Additionally, it includes a review of how the Rift Lakes were formed and how the cichlids in them evolved. There is also a recount of an eventful collecting expedition to Lake Tanganyika. Thus this book is intended to introduce Rift Lake cichlids to the beginner and intermediate hobbyist alike, and it is also intended to interest the advanced hobbyist in some new ideas and interesting information. Additional reading is listed in the back of this book.

The author with his Tanzanian diving team on Lake Tanganyika. The Tanzanians were trained by Misha Fainzilber, Tanzania's first fish exporter.

Lamprologus brichardi and other Tanganyikan cichlids amongst the rocks (photo by Glen S. Axelrod).

Introduction

The cichlids of East Africa are renowned in science for their proliferation and diversity and in the aquarium hobby for their beauty and behavior. Most of these fishes come from lakes that formed when two great valleys filled with water millions of years ago. Often thought of as only the huge Lakes Tanganyika and Malawi, the African Rift Lakes also include Lakes Kivu, Edward, Albert, Rudolf and a number of other relatively small lakes and ponds. This book will principally deal, however, with two of Africa's Great Lakes, Malawi and Tanganyika. Lake Victoria, an extensive shallow saucer-shaped lake, is a depression between the two East African Rift Valleys and is the other Great Lake of Africa.

The Great Lakes of Africa are unique in that their vast cichlid faunas are principally endemic or, in other words, found within their respective lakes and nowhere else. In all, Lake Malawi has about 285 fish species of which about 217 are cichlids, and Lake Tanganyika has about 210 fish species of which about 140 are cichlids. This helps account for the fact that Africa has more cichlids than any other continent.

African cichlids are sought in the aquarium hobby for both their beautiful coloration and remarkable behavior. It is often said that no other animals, except birds, can offer the brilliance of coloration that is found in fishes. Rift Lake cichlids are not only among the most colorful of freshwater fishes but also among the most complex and ritualistic in their behavior.

9

An aerial view of Lake Tanganyika just south of Kigoma (photo by Glen S. Axelrod). *Julidochromis dickfeldi* is a relatively peaceful Tanganyikan cichlid which exhibits a graceful beauty in a Rift Lake aquarium (photo by Dr. Herbert R. Axelrod).

Rift Lake Cichlids and Rift Lake Cichlid Evolution

The Rift Lakes

Lakes Tanganyika and Malawi were formed as a result of tremendous earth movements. The Great Rift Valleys which contain the lakes began to develop in Miocene times (15 million years ago). The eastern Malawi Rift Valley and the western Tanganyikan Rift Valley were never connected. Lake Tanganyika consists of two continuous troughs that significantly altered the drainage system of the area as they began to form. The Malagarasi River was cut in two by the Tanganyikan Rift, and the upper reaches of the river (to the lake's east) drained directly into the huge crevice. The lower portion of the river (to the lake's west), which is probably represented today by the Lukuga River, was cut off from the supply. This internal drainage could have lasted as long as ten million years.

Lake Tanganyika was isolated for six million years and as recently as 100,000 years ago gained its outlet to the Congo River System. This event was stimulated by changes caused by volcanic activity that occurred north of the lake. Prior to this activity, Lake Edward and the Nile River System served as the receptacle for drainage from the Ruanda Highlands. About one million years ago, however, the Bufumbiro Volcanoes formed across the northern part of the rift valley and impeded the drainage to Lake Edward; Lake Kivu formed against these volcanoes, eventually spilling over to the south. This overflow formed the Ruzizi River, which flowed through the Ruanda Highlands and into the north of Lake Tanganyika. The Panzi Falls separated the fishes of Lakes Kivu and Tanganyika. Lake Tanganyika itself began to rise. It eventually overflowed through the old western portion of the Malagarasi River, which is now called the Lukuga River, thus ending Lake Tanganyika's long period of isolation.

Today, Lake Tanganyika is 650 km (404 mi.) long, has a maximum width of 80 km (50 mi.) and lies at an altitude of 773 m (2536 ft.). The lake has a surface area of 34,000 km² (13,100 sq. mi.), almost twice the area of the state of New Jersey. The main inflows into Lake Tanganyika are the Ruzizi and the Malagarasi Rivers. The main outflow is the Lukuga River, which is occasionally blocked by accumulated swamp vegetation. Evaporation accounts for 95% of the total water loss from the lake, as the annual outflow is only 1/1500 of the lake's total volume. The mineral-salt concentration, although higher than in the rivers, is 420 parts per million (ppm). The pH (measure of acidity or alkalinity) of the lake is alkaline and in excess of 8.6, reaching 9.3. The lake is the deepest African Rift Valley lake and the second deepest in the world after Lake Baikal in the U.S.S.R. A depth of 704 m (2300 ft.) has been recorded for the northern basin (lat. 5 °S.) and a depth of 1470 m (4820 ft.) for the southern basin (lat. 7 °S.). Because of deoxygenation and the retention of hydrogen sulphide (chemically poisonous to fish) at greater depths, the fish are contained in the uppermost 100 m (330 ft.). This oxychemocline (barrier caused by oxygen and chemical changes) ranges between depths of 40 and 100 m (130 and 330 ft.). Its existence results from the stable climatic conditions of the region, since temperature stability in the lake (24-26.5 °C or 75-80 °F) prevents thermo-currents (water movement due to temperature differences) that would otherwise create a circulation. This circulation would oxygenate the deep water and eliminate hydrogen sulphide. Since this is not the case, only a relatively thin surface layer of water sustains aerobic (oxygen respiring) life in Lake Tanganyika.

The western African rift axis contains several lakes, the largest being

Lake Malawi. Lake Malawi has the same general appearance and form as Lake Tanganyika. It is the ninth largest and fourth deepest lake in the world and the second deepest African Rift Valley Lake. It is also second to Lake Tanganyika in age (for African Rift Valley Lakes), being one to two million years old. Geological evidence for Lake Malawi is even more scarce than for Lake Tanganyika, but it is postulated that both lakes formed along similar lines. Lake Malawi has only one deep basin, which is in the northern end of the lake—and it is here that the lake originated. Its present shape and level developed only after many years of complex earth movements. Like Lake Tanganyika, the Malawi Rift probably intersected a large river that helped to fill the lake. That river was probably the South Rukuru which, most likely, was once a part of the Ruvuma River System, emptying into the Indian Ocean to the east. Lake Malawi gradually extended southwards from its northern basin. Although Lake Malawi once extended farther south than it does today, it was always isolated by the Murchison Rapids. These rapids effectively prevent animals living in the lake from passing to the Zambezi via the Shire River.

Today, Lake Malawi is 560 km (350 mi.) long, has a maximum width of 80 km (50 mi.) and a surface area of 30,000 km² (11,600 sq. mi.); it is 457 m (1500 ft.) above sea level. The catchment area of the Malawi basin is very small compared to that of Lake Tanganyika, and the inflows into the lake are mostly short streams from the immediate surrounding area. The main outflow is the Shire River to the Zambezi. The Shire, the "Malawian" Lukuga River, is intermittently blocked by accumulated swamp vegetation and changing river levels. This blockage, however, is far less frequent than that for Lake Tanganyika, as there has only been one blockage cycle in the past century. Like Lake Tanganyika, Lake Malawi also has a high mineral-salt concentration and a high pH (7.7-8.7). Unlike Lake Tanganyika, Lake Malawi has only one deep basin, which has a depth of 770 m (2500 ft.). Lake Malawi has little temperature variation; the range between the surface and bottom waters is only 3.5 °C, from 24 to 27.5 °C (75-82 °F). Thus, both Lakes Malawi and Tanganyika are permanently stratified, meaning that there is negligible water turnover. Lake Malawi is perpetually deoxygenated at a depth greater than 400 m (1300 ft.).

The results of Boehm's 1883 expedition to Lake Tanganyika led to the first theory of the lake's origin. It was postulated that the lake was once connected to the Atlantic Ocean across the Congo River Basin. This theory was based upon the incidence of fauna that appeared to be of marine origin. It gained wide support among prominent biologists of the time, notably Guenther and Moore. The marine-like forms are now

Highly specialized behavior has evolved in Rift Lake cichlids. (1) *Haplochromis polystigma* male becomes lighter in color when aggressive and is seen here attacking its image (photo by Glen S. Axelrod). (2) *Lamprologus tetracanthus* female aggressively guards her young (photo by Glen S. Axelrod). (3) *Lamprologus brichardi* has elaborate breeding behavior (photo by H.J. Richter).

2

known to be related to other African freshwater fauna. In 1920 Cunnington discredited the "sea theory," and no geological evidence has been found to support it.

The transection of the Malagarasi River by Lake Tanganyika has been supported by the presence of several species of fishes that are not found in the lake but occur in the Malagarasi and Congo Rivers. These two rivers are completely separated by Lake Tanganyika. Several (noncichlid) examples of such species are: *Labeo weeksii, Barbus eutaenia, Polypterus ornatipinnis, P. congicus, Distichodus maculatus* and *Tetraodon mbu.* These fishes are assumed to represent early riverine forms that were geographically separated during the lake's formation. Although these fishes did not enter the lake, other fishes from the Malagarasi River did.

Rift Lake Cichlid Evolution

Darwin's (1859) observations on the finches of the Galapagos Islands offered the first example of explosive speciation to be recognized as such and studied in depth. Since that time many other striking cases of this phenomenon have been uncovered and examined. Perhaps the most significant example of explosive speciation can be found in the Great Lakes of Africa. Here the fish faunas are dominated by one large family, Cichlidae. The cichlids have evolved within the individual lakes to a point where hundreds of endemic species now exploit almost every conceivable way of life. Lake habitats successfully colonized by cichlids include rocky shores, sandy shores, pelagic (open water) regions and bathypelagic (deep open water) regions. Lake Tanganyikan cichlids offer the most outstanding example of intralacustrine (species formation within the lake) speciation and species diversification. To date 40 genera have been described, of which 34 are considered endemic. Furthermore, all of the lake's (approximately) 140 known species are considered to be endemic. Lake Malawi also offers an outstanding example of intralacustrine speciation, although its cichlid species diversification is not so extensive as that in Lake Tanganyika. Lake Malawi contains 26 different cichlid genera, of which 22 are considered endemic. Additionally, almost all of the lake's 217 known species are considered to be endemic.

INVASION

Little is known about African fresh waters before Miocene times (15+ million years ago). The invasion of Lake Tanganyika by fish from the Malagarasi River is assumed to have immediately followed the initial formation of the Rift Valley (during Miocene times). As in Lake Malawi, Lake Tanganyika's dominant fish family, Cichlidae, is assumed to have arisen from the ancestors of the generalized riverine forms of *Tilapia* Smith 1840 and *Haplochromis* Hilgendorf 1888. *H. bloyeti* has the characteristics one would expect from riverine ancestors. It has a generalized anatomy, is relatively small (usually less than 14 cm. [5½ in.] in standard length), omnivorous but leaning toward a largely unspecialized carnivorous diet and can tolerate a variety of conditions.

ADAPTATION

It is a widely recognized biological postulate that, in order for animals to survive, they must be adapted to the environment in which they live. Adaptations in a changing environment are always inseparably coupled with function and, particularly in the present case, are largely structural (body) or behavioral modifications. Adaptive radiation refers to the territorial advancement of an animal that will then adapt to the changing environment it encounters. It involves behavior, habits and physiology (mechanisms by which living things function), as well as morphology (form and structure). The invading riverine fishes entered a lacustrine habitat that is composed of three basic regions. The littoral region is comprised of the bottom reaches of the lake and the water above these bottom reaches where it is clear enough and shallow enough to allow photosynthetic rooted flora to exist. Thus, this region is comprised of the shore area and is bounded by the point at which there is not enough light to support rooted green plants. In Lakes Malawi and Tanganyika, because of the steep gradient of descent, this region is found only along the shoreline. The region is composed of rocky outcrops and sandy shores, and it contains the majority of the two lakes' faunal diversification. The benthic region is the area of the lake bed below the littoral region and bounded in its lower limit by the oxychemocline (the line which marks the area of perpetual deoxygenation discussed earlier). The pelagic region is that area of water over the benthic region and the oxychemocline, and bounded by the littoral region.

FROM RIVER TO LAKE

African fishes must have passed through certain stages of evolution before they could adapt to the lacustrine environment. The initial primitive condition is a complete riverine existence that entails feeding and breeding only in the rivers. As the fishes begin to evolve toward a lacustrine existence, they feed in the lakes and rivers but breed only in the rivers. The height of generalization comes in the third stage, in

1

2

Pseudotropheus zebra
color morphs: (1) white,
(2) tangerine, (3) cobalt
blue, (4) mottled OB, (5)
mottled, (6) cobalt
green. Photos 1 & 6 by
G. Meola, African Fish
Imports, 2 by Kochetov,
3 & 5 by Dr. Herbert R.
Axelrod, 4 by Dr. Warren
E. Burgess.

3

4

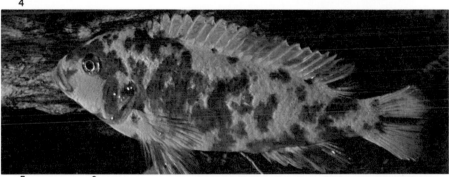

5 6

which the fishes feed and breed in both the lakes and rivers. Finally, in the fourth (and last) stage, the fishes become specialized lacustrine inhabitants by feeding and breeding only in the lakes. Today, almost all of the cichlid fishes within the Great Lakes of Africa are confined to their particular body of water and barred from riverine life because of their specializations. Over the course of time, geographical changes in Lakes Malawi and Tanganyika have resulted from such events as changes in the lake level, rock slides, silting from inflowing rivers and erosion of rock surfaces. One result of these changes has been to create new barriers and remove old ones. This process created many isolated areas or "pockets of fish life" called micro-habitats. The term "microgeographic isolation" refers to the fish isolated in these pockets which are within one lake. Thus, separate development took place within these isolated areas.

THE EXAMPLE OF *TROPHEUS MOORII*

Tropheus moorii is an example of an endemic cichlid widely distributed in Lake Tanganyika which is undergoing speciation through microgeographic isolation. Records of *T. moorii's* distribution on the western shores of the lake are imprecise at the moment and are not cited here. In three expeditions during 1976 and 1977, I examined the distribution of *T. moorii* in the eastern central part of the lake. Like the northern portion, this part consists of rocky outcrops separated by long stretches of sandy beach. *T. moorii* is a rock-dwelling fish that is rigorously restricted to its habitat of rocky outcrops and reefs along the littoral zone. It never moves more than one meter (3 + ft.) away from the rock that it inhabits. Thus the sandy beaches act as geographical barriers. There are many known color morphs of this fish. The morphometrics of these fish are virtually the same, but different color varieties might be regarded as subtaxa, as they are reproductively isolated (geographically) and there are no color intermediates. Here is an excellent example of differentiation in its early stages. The different populations are isolated from one another and are morphologically diverging (assuming color to be a morphological characteristic). They can still interbreed, however, and this is illustrated by the Kashikezi population, which appears to be intermediate between the Luhanga and Bemba groups. Although many of these populations may never reach species status, one probably already has. *Tropheus duboisi* seems to have been a color morph of *T. moorii* that has diverged to the point at which it may be regarded as a distinct species. It co-exists with the orange form of *T. moorii* but is reproductively isolated.

A view of Lake Malawi from Cape MacClear, Malawi. The natives fish in small dug-out boats for the open water fishes of the lake (photo by Dr. Herbert R. Axelrod).

PATHWAYS OF EVOLUTION

The pathways of evolution from the ancestral stock (or stocks) of Africa's Great Lakes to the present day fish fauna remain unclear. This results, in part, from the fact that the fishes have undergone many generations of adaptive radiation. The present day fishes have evolved to the point that their physical and behavioral specializations are far different from their ancestors'. Several mechanisms such as parallel and convergent evolution complicate phyletic studies.

Boulenger (1906-1916) was the first to arrange taxonomically the entire known Tanganyikan and Malawian fish fauna and place his conclusions within one work. Since that time, much work has been done in an attempt to elucidate the ancestral relationships of the formulated species. Analysis of the situation has been based almost solely on external morphology. Trophic specializations and behavioral characteristics have only recently been taken into consideration. Deductions from evidence about these modes of analysis have led to many conflicting hypotheses. It is clear, therefore, that considerably more research is needed in this area to help clarify many of the problems and contradictions in African cichlid taxonomy and phylogeny.

The steep face of "Mgu wa Tembo," which is translated from Swahili as "Elephant's Foot," is a well known landmark for the fishermen from Kigoma Bay on Lake Tanganyika (photo by Glen S. Axelrod). *Trematocranus jacobfreibergi,* a rock-dwelling cichlid from Lake Malawi, displays the beautiful coloration for which many of the Rift Lake cichlids are well known (photo by H.J. Richter).

Collecting On The Lakes

Collecting work is an important part of a scientist's research. The best way to study a fish's ecology and biology is in the field. Here one is able to observe the fish in its natural habitat and determine the best conditions under which it should live in the aquarium. During the past three years, I have been fortunate in that I have been able to make three expeditions to Lake Tanganyika and two to Lake Malawi. It is impossible to relate the events of the entire series of trips. My last trip to Lake Tanganyika, however, was a bit out of the ordinary and is worth recounting.

It was six o'clock in the morning when my plane left Dar Es Salaam for Kigoma, a Tanzanian harbor town on the eastern shore of Lake Tanganyika. The yellowing appearance of the countryside as the old plane passed over the expanse of central Tanzania was caused by the fact that it was September and the dry season was well under way. This was my third trip to Tanzania and my second to the lake itself. My previous visits were to lay the groundwork for this longer expedition. I was fortunate to be working with Misha Fainzilber, the Tanzanian fish exporter whom I had met during one of my previous visits to Tanzania. As the plane descended, Lake Tanganyika became visible on the horizon; my work had suddenly begun.

I was met at Kigoma by Misha's nephew and operations manager Isaac Abrahamowitz. At this point I ran across my first problem and challenge. ALL OF MY BAGGAGE WAS LOST!!!! Apparently, for one reason or another, my baggage was not put on the plane in Dar Es Salaam. Fortunately, I had taken the necessary precaution of carrying all of my essential scientific and camera equipment onto the plane with me. As things turned out, I was never to see my baggage again. Thus I was to live a more "natural" existence over the four week trip than I had previously planned.

After a day's rest, Isaac and I went down to the fish station on the lake. The collecting operation was quite impressive and had come a long way since my last Tanganyikan visit four months earlier. Isaac had two six-man teams of divers and three power boats. During my previous trip to the lake, Isaac had had only one boat working. I vividly remember the two of us taking the boat out one afternoon and having the engine quit on us. As it turned out, we ended up rowing for four hours to get back to Kigoma—after three hours of diving. Fortunately, things seemed to be more organized this time. Now when the divers come back with the day's catch, the fish are placed in an area of the lake that is enclosed by a screen. This fish-keeping technique eliminates the problems of water changing, aeration, and temperature control that were so difficult to cope with before.

One week after my arrival in Kigoma, Misha came up from Dar Es Salaam and had a new boat shipped up by train. The new boat was much faster than the others. It was a fiberglass speedboat powered by two 48 HP Yamaha engines. It was capable of doing 45 knots and enabled expeditions farther afield than were previously possible with Misha's slower boats.

Up to this point, I had been going out with the divers in the morning and returning at about noon. Because of the expense and time involved in more extensive trips, all of our work until now had been done within a twenty-five mile radius of Kigoma. With the new boat on hand, that had changed. Misha had arranged for us to make a long expedition to Miyako and Kasoje in the central portion of the lake.

Two days before our departure, we began to collect supplies and plan for the expedition to the south. We would be in either uninhabited or socially primitive regions, and it was therefore important that the trip be carefully planned. I was told by some fisherman that the expedition would take us through areas inhabited solely by cannibals! Fortunately, however, according to the fisherman, these tribes live in the mountains surrounding the central portion of the lake and believe it to be bad luck to either come into contact with or see the lake. Thus, as it would be difficult to find provisions during the expedition, we collected some canned food, cassava roots and dried fish to eat during the five day trip. We had to leave early in the morning or late at night, as this was the time that the lake was usually calmest. A long trip on even a moderately rough lake would be unbearable. The boat was small, about the size of a large automobile, and we would be traveling approximately 50 miles per hour!

Thursday morning we got up at 2:00 AM to begin the trip south. There were eight divers, including Isaac and myself. We all met at the boat, which had been provisioned and packed the night before. We were about to leave when we got our first bit of bad luck. The gasoline had not yet arrived at our first refueling stop. The water taxi carrying the gas had left for its destination eight hours late. (A water taxi is a slow-moving lake boat that is used to carry both passengers and cargo.) Thus we had to wait until late evening before we could leave.

We began our journey at dusk and quickly sped far out onto the lake. As we left Kigoma Bay, we rounded the great cliffs of the "Mgu wa Tembo," which is translated from Swahili as "Elephant's Foot." The steep rock face resembles a great elephant's foot from the lake at a distance of a mile or more. But because the evening was cloudy, it became dark quickly, making visibility poor. We carefully went over our plans and then settled down for the long trip south. As we passed Ras Meno (Meno Point) and Ras Kitwe (Kitwe Point), we entered Bangwe Bay and could see the fishing boats coming out from Ujiji harbor, preparing for the night's fishing. Almost all of the food fishing along this part of the lake is done at night. Boat lights are used to attract the fish, which are caught in large seine-type nets. Farther south, however, most of the fishing is done by day, using more primitive fishing methods and canoes that are dug out from logs. Isaac's divers fish Bangwe Bay for a yellow morph of *Tropheus moorii*. *Tropheus duboisi*, *Lamprologus tretocephalus*, *Spathodus irsacae* and *Cyphotilapia*

1

2

(1) Miyako collection location.
(2) *Xenotilapia longispinis*
spawning in Lake Tanganyika.
(3) *Aulonocranus dewindtii* (4)
Kigoma area. (5) *Lamprologus
tretocephalus* guarding young
in Lake Tanganyika. (6) *L.
tretocephalus.* Photos 1-5 by
Glen S. Axelrod, 6 by D.
Scheurman.

3

4

5   6

frontosa are also caught here but can also be found in many other parts of the lake. Most of these fishes do not have local names, because few are considered edible by the local people. *Cyphotilapia frontosa* is an exception. Called "Ndubu" by the locals, it can easily be caught with bait from a hook and line, and it is considered almost a delicacy.

It was unfortunate that we were traveling at night, as I was looking forward to seeing more of the beautiful Tanzanian countryside. Although I couldn't take pictures of the landscape at night, I would not have taken them during the day either. At various points along the lake, the taking of photos is forbidden because of the presence of military installations. These installations were recently set up and inspired by Tanzania's concern over terrorism. Three years earlier, several foreign students were kidnapped by Zaire terrorists and taken across the lake. Although the government of Zaire managed to obtain the release of the hostages, Tanzania is unwilling to take any chances that might jeopardize the safety of her people and visitors. The bases are for protection, and photographs of the area would be a security risk.

It was pitch black as we entered the Malagarasi delta region. We were speeding along at about 50 miles per hour on an extremely calm lake when I was suddenly thrown from the boat. When I hit the water it felt as if I had hit a slab of concrete! As I stood up in the water and collected my senses, I realized that we must have strayed too close to the shore and had run into the Malagarasi River delta, which reaches out several miles beyond the river mouth. The water was only two feet deep! The propellor of the left engine had hit the sand and turned the boat sharply, throwing me out and lifting the entire engine into the boat. Fortunately, no one was hurt. As I made my way back to the boat, I realized that it was a good thing that I had been thrown out. If I had remained in the boat, I would have been crushed by the engine, which was now lying where I had been sitting. Six months earlier I had encountered a close call of a different nature in this area. We were skin diving for *Polypterus,* a valuable fish for palaeontological and taxonomic study because of its primitive condition. As we were about to leave, a hippopotamus began to charge toward us. She was several hundred feet away but was nevertheless upset by our presence. We quickly started the engine of our boat and were off before she was even close. It was a scary experience that I vividly remembered, so I now strained my ears to listen for a hippo's familiar snorts. After an hour of repair and two hours of rowing our way out of the labyrinth of sand bars, we were on our way again.

At a village near Kiti Point, we stopped to drop off two divers and equipment. They would concentrate on catching a green morph of

Tropheus moorii, a yellow morph of *Lamprologus compressiceps,* and a yellow morph of *Julidochromis regani.* We also decided to stop here for a few hours and get some sleep, using our sleeping bags on the beach. Compared to the scorching heat of the day, it was now extremely cold, and I fell asleep in my wetsuit.

I awoke at about 8:00 AM surrounded by a score of curious natives, so I got out of my sleeping bag, which was becoming quite hot. Feeling a strange pricking on my stomach, I zipped open my wetsuit to find that I was covered with thousands of black ants. The natives had a good laugh as I shot for the water, stripping along the way. This beach was used by the villagers to dry their fish. The ants cannot survive the scorching daytime sun, so they hide where there is shade and come out onto the beach at night. As the morning sun warmed the sand, my body and sleeping bag provided an excellent refuge for them.

Several dozen native fishing boats saw us off as we sped toward our first refueling stop. The surrounding countryside was mountainous and carpeted with an intensely green and thick jungle growth. During the previous night, we had moved from the *bundu* (Swahili for "bush country") to the jungle country. We soon arrived at our first stop of the day, the small village of Mkuyu, nestled against the slopes of a steep hill that arced around a protective bay. Dozens of villagers ran out to greet us with gestures of friendship and hospitality. One of our divers came from this village and here he was considered to be a very important man. He spoke to the chief of the village in a local language that was different from the familiar Swahili. Several other divers collected the gasoline that had arrived the previous night by water taxi. An arrangement was made with the chief to use his village as a permanent "gas station." For this privilege Isaac paid the equivalent of 25 American dollars—a small fortune here that would be used to build three or four new houses.

Fully refueled, we resumed our journey south; since the water was calm, we made great time. Isaac pointed out Miaba and Halembe to me. Miaba was a great place to catch *Lamprologus leleupi* as well as a morph of *Tropheus duboisi* that has olive bands. Through the haze hanging in the distance we could see the mountains above Bulu Point. As we approached the point I noticed that there was little or no beach, the mountains dropping directly into the water. From a distance the mountains looked like huge moss covered hills that were shrouded in the mist of the dry season. At their base lay rocky outcrops that shelter scores of different cichlid species, many of which are still undescribed. We passed Bulu Point and were again within clear sight of land. The only signs of human habitation that I could see were the small fires in the mountains.

1

2

Tropheus duboisi color morphs. (1) broad yellow-white banded; (5) is juvenile of same morph. *T. moorii* color morphs: (2) yellow-belly, (3) *T.m. kasabae,* (4) blue, (6) orange-sided. Photos 1, 5 and 6 by P. Brichard; 2 by Glen Axelrod; 3 and 4 by W. Staeck.

3

4

5   6

We arrived at our destination late in the morning. The campsite had already been established by a group of Japanese researchers studying the behavior of chimpanzees. Our boat was unloaded, and we hurried out to get some diving done before lunch. We broke up into two teams. We dropped off Isaac's team and then moved on to another locality, Bulu Island. Because we were free-diving, our work was confined to a maximum depth of 40 feet. We anchored our boat well away from the rocky shore. Then a 30-foot net about one yard high was stretched across the rocks on the lake bed. Because of its transparency, most of the fishes would be unaware of the net until they were chased into it by a diver. As the fishes tried to force their way through the large net, the diver caught them in a small hand net. From the hand net, the fishes were placed in submerged or floating barrels with self-closing elastic tops. The barrels had hundreds of small holes drilled halfway up their sides so as to provide as much water exchange as possible.

After two hours of diving, we headed back to camp, picking up Isaac's team along the way. We were quite excited about the catch. Isaac was successful in catching a tremendous number of *Chalinochromis brichardi* and *Lamprologus leleupi*, while my team was able to boast of a hundred-plus red *Tropheus moorii*. In addition, I was very excited about some new fishes. Two of the potentially new fish resembled *Chalinochromis brichardi*. A third fish, *Tropheus polli*, was described by the author (J.L.B. Smith Institute of Ichthyology Special Publication 17, November 1977).

When we arrived back at camp, the Japanese research team headed by Dr. Itani, a zoologist and animal behaviorist, was waiting for us. Dr. Itani has been studying primate biology in Tanzania for the past 15 years. We all had lunch and then, at about 2:00 PM, I started photographing the fishes we had caught that morning. I had photographed about 25 fishes when I was unexpectedly distracted.

I was snacking on a banana while arranging the final fish in my photo tank. While concentrating upon the fish, I put the banana down on the table next to me and out of the corner of my eye saw a hairy arm grab my snack. As I turned in astonishment, the hairy arm and the 100-pound chimpanzee attached to it quickly made off for the nearest

(1) Malagarasi River Delta swamp. (2-6) Dr. Itani's chimpanzee research station proved to be more entertaining than expected when the chimps came down from the hills. (2) My research area, (3) the chimp in the tree who stole my banana, (4-6) feeding the chimps (photos by Glen S. Axelrod).

tree. As I turned back to my work, I noticed another chimp fancying my camera and equipment. I quickly packed up all of my equipment as a dozen more chimps came down from the hills. I soon realized why. For the past couple of years, Dr. Itani has been encouraging the animals to come down to the camp by leaving out food such as bananas and sugar cane. As time passed, the chimps became brave enough to take food directly from the hands of the Tanzanians working for Dr. Itani. Chimpanzees are extremely powerful for their size. Furthermore, this large African ape (which is related to the gorilla) can weigh as much as a man and possesses impressive canine teeth that are several inches long. So with caution in mind, I put my work aside and joined the Tanzanians in the feeding. Soon the entire tribe of chimpanzees was feeding. Their expressions and reactions were amazingly human-like.

That evening Isaac and I were invited to a celebration given by the local people of Kasoje in honor of Dr. Itani, who was leaving the next day for Kigoma and then taking a trip back to Japan. The ceremony was an exciting display of singing and dancing. First, however, we feasted on a special dinner prepared by Dr. Itani himself. It was a Japanese meal prepared from chicken (a rare and expensive commodity in this part of the world) and vegetables. Then the local people provided a feast of mashed cassava roots and dried fish. The fish were boiled in water. The cassava roots, which looked like mashed potatoes, were taken from a communal pot with the left hand and molded into the shape of a small bowl. The fish was pushed into the center of the cassava mold and eaten in that form. The method and manner in which the preparation was made and eaten is an important part of tribal etiquette. For example, the right hand is never used to eat with, since it normally takes the place of toilet tissue in this part of the world. Thus it would be considered extremely impolite to place the right hand into the communal pot!

After the feast, the local people began to sing and dance. Isaac and I were privileged to sit at the guest of honor table with Dr. Itani and his colleagues. Each song and dance was rich in tradition and carried an important story or moral. After half an hour, some of the tribesmen came up to the table and took each of us in turn to dance with them. It was a real challenge as I tried to mimic the natives' steps and style. As the celebrations, which lasted through the night and into the next morning, came to an end, there was suddenly a great deal of nervous commotion. Moving from the village to the beach, I gazed toward the mountainous south and was awed by a tremendous fire. It was still far away but stretched over several miles, leaving a charred trail in its wake. In several days it could reach us, but we were planning to leave for Kigoma in two days.

The next morning we went south for another series of dives. The previous day's catch was kept in the perforated plastic drums and anchored off the shore at about a 15-foot depth. We proceeded toward Pasagulu Point in search of the red-spotted morph of *Tropheus moorii* and the striped and spotted morphs of *Chalinochromis brichardi*. These fishes are found in various locations from Pasagulu Point south to Edith Bay. The red-spotted *T. moorii* differs somewhat from the red *T. moorii* found at Bulu Point and Bulu Island. Again we noted the huge fire along the mountains; it seemed much closer to us. We dove for about three hours. While the other divers concentrated on *Tropheus* and *Chalinochromis,* I collected specimens of as many different species as possible, then photographed them in the afternoon. The next day was our last working day, so I tried to make the best of it. I went out with the divers in the morning and afternoon. During the afternoon dive it began to rain, making the fishes in the shallow water extremely active and quite a bit more difficult to catch. As I was chasing one interesting specimen around a rock, trying to coax it toward the net, I came face to face with a large Tanzanian water cobra *(Boulengerina annulata)* about two meters in front of me. Its hood was open, its body was coiled and it appeared ready to strike! My initial reaction, one of scientific curiosity, was closely followed by fear. It is an extremely venomous snake whose bite could be fatal without immediate medical attention. As I slowly retreated, another diver came up behind me, not seeing the snake. Fortunately though, the snake decided to swim away. I later learned that this is not an aggressive snake; it will usually retreat rather than attack.

We left Kigoma the next morning, picked up the other team of divers at Kiti Point and arrived back in Kigoma just before noon. The five-day expedition was the highlight of my four-week trip to Tanzania and Lake Tanganyika. The entire Tanzanian trip had been a success, as I had the opportunity to take many fine photos and collect several new fishes in addition to collecting a wealth of ecological information on cichlids. The trip was necessary for my research into Lake Tanganyikan cichlids and would not have been possible without the help and cooperation of Misha and Isaac. I owe a debt of gratitude to both of them for their unending generosity, and to the Tanzanian people for their hospitality.

A *Melanochromis auratus* male digs the gravel out from under a rock in order to build a nest (upper photo by Dr. Herbert R. Axelrod). A pair of *Genyochromis mento,* between the wood, are rough customers in an aquarium. In addition to harassing other fish, they will also try to rearrange your aquarium to their liking (lower photo by A. Ivanoff).

Setting Up And Maintaining Your Aquarium

Generally speaking, African Rift Lake cichlids are very hardy and will withstand a wide range of conditions. Nevertheless, conditions simulating those of the Rift Valley Lakes are necessary if you desire to spawn your fish, want them to live long and want them to exhibit optimum coloration. Because of their size and territorial nature, most African Rift Lake cichlids require medium to large aquaria. Most grow to at least 10 cm (4 inches), and when there is more than one mating pair in an aquarium, their territorial nature will lead to confrontation. The less dominant fish could be, and most probably would be, killed if it had no room for escape. Keeping this factor in mind, one must choose the type of aquarium community desired before proceeding with material purchases. The beginner should gain experience with one pair of cichlids before attempting to handle a larger array.

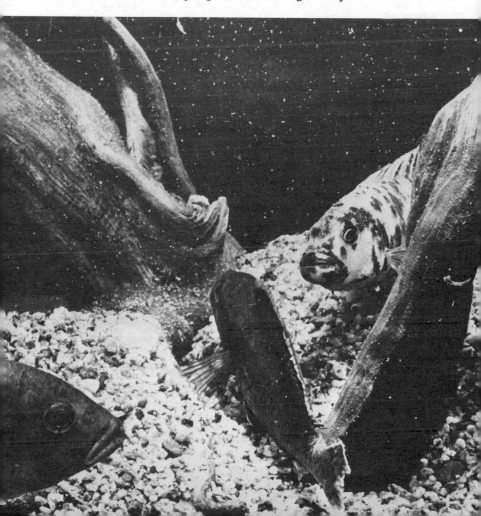

PLANNING

A 20-gallon tank is a good size to start with, although a 30-gallon tank would allow more room for expansion at a later date. Start with a dry aquarium and sketch your layout before you proceed. It is important to note that Rift Lake fishes are constant diggers. They will dig up almost any plant without exception and will uncover undergravel filters and will undermine rock arrangements if not properly constructed. This can lead to more problems than expected. Valued plants will be killed (uprooted and then eaten), biological filters rendered inactive, and cherished fishes could be crushed in collapsing caves. Furthermore, if undermined rock piles are high enough, they can fall and break the aquarium glass.

ROCKS

Most of the African cichlids available for sale are rock dwellers. This group includes the Malawian mbunas and the Tanganyikan genera *Tropheus, Simochromis, Petrochromis, Limnochromis, Ectodus, Julidochromis* and some *Lamprologus*. These fish not only live among the rocks for protection but also spawn on and under them and eat algae, insects and crustaceans off them. Hence a good number of rocks forming caves and outcrops in the aquarium will help to create a natural atmosphere for your fish. The rocks should be placed directly on the bottom of the aquarium, with the gravel around them. If gravel is placed under your rock formations, the fish will undermine the foundations. Flowerpots can also be used as caves. Care should be taken to be sure that your rocks and gravel do not contain any metal or poisonous material soluble in water. Wash the rocks and gravel very carefully before placing them in the aquarium. One method of arrangement involves gluing small rocks together using silicone cement. This will make chunks of rock that you can arrange again but that are too big for your fish to rearrange. The advantage of this method is that you can rearrange the aquarium to give it a completely new look should you desire to do so. The second method of arrangement involves gluing the big rocks together with silicone cement and leaving the small rocks unglued. A drawback of this method is that you are unable to easily rearrange the entire aquarium. You would first have to cut the big rocks apart at the points at which they were glued. You can, however, still change the cave structure by using the small stones to block off some caves and open up others, creating a new environment for your fishes. This second method also has the advantage of eliminating partial rock cave-ins caused by cichlid tunneling. All of the big rocks are connected into one unit.

Rockscaping in your aquarium is a challenging necessity. Without some creative ingenuity on the part of the Rift Lake aquarist, his aquarium, devoid of plants, will appear barren. It is as important to create a frame around your beautiful cichlids as it is to create a frame around a picture.

There are several factors to take into consideration when rockscaping your aquarium. Consider the center of your aquarium the focal point. This point draws the eye, and everything else revolves around it. This area should be open and uncluttered. Your rockwork should spread out from this point. If you want to build a cliff, create one in the back of the aquarium or on a side of the aquarium that is difficult to view.

The rockwork architect has many variables to work with. Consider the size and shape of each rock. It is important to lay your rocks out so that you have some degree of continuity. An arrangement that is repetitious or too variable will not be pleasing to the eye. You must use your rocks' varying sizes and shapes to create transitions lending to the continuity. Avoid sharp contrasts. A random flow of rockwork is usually most successful, as it more closely approximates nature. Personally, I try to copy (on a miniature scale, of course) scenes that I have seen in the lake.

Rooted plants are not recommended for the cichlid aquarium because they will be uprooted and eaten. Some people turn to plastic plants, which are safe and quite acceptable. I, however, would rather rely on the beauty of my rock formations for esthetics. I find that having plastic plants around my prize cichlids is analogous to placing beautiful gems in iron settings.

AQUARIUM pH

The pH (measure of relative acidity or alkalinity) of your aquarium water is important to monitor and necessary to keep high (alkaline). Lake Malawi has a pH range of 7.7-8.7, and Lake Tanganyika gets as high as 9.3. For this reason, it is best to keep your aquarium pH at about 8.5. Hard water will help, and you should use an alkaline-base gravel such as marble chips as a buffer. Special mineral salts are also available. Personally, I have had good success with a mixture of 7 parts tap water with 1 part sea water. Most Rift Lake cichlids will survive at a neutral pH (7) or even at an acid pH of 6. However, experience has shown that as the pH falls below 8, the fishes tend to become more susceptible to disease, lose their coloration and cease to breed.

FILTRATION

Filtration is a very important consideration in any aquarium system.

There are two basic kinds of filtration systems used in aquaria: the particulate filter and the biological filter. The particulate filter does just what its name implies; it removes particles of dirt and debris from the aquarium, usually trapping them in filter wool. This filter can either be placed within the aquarium or, preferably, hung outside the aquarium. There are many particulate filters to choose from at varying prices. Some of the better and more expensive particulate filters use diatomaceous earth rather than filter wool as a filter medium. The diatomaceous earth is composed of hard silica skeletons of dead microscopic water plants called diatoms. Your aquarium dealer can help you choose the best particulate filter for your aquarium, depending on its size and its biomass (amount of living material within your tank).

A biological filter is essential in almost any well run aquarium and can be used in conjunction with a particulate filter, which is optional. The most common biological filter is the undergravel filter. Slightly raised plastic supports are placed under the gravel; they have small holes or slits in them, openings small enough to prevent the gravel from falling between them but large enough to allow the water to circulate. Water is drawn through the gravel and plastic gravel support plates and passed (pumped or bubbled) up airlifts to the water's surface. In this case, the gravel acts as the screening agent and also as a biological filter. The gravel contains a number of different microorganisms that digest and break down decaying food matter, decaying plant matter and fish excrement. Matter that is normally toxic to your fishes is thus broken down into mostly harmless components. Additionally, this filtration system prevents a build-up of anaerobic (without oxygen) activity in the gravel that could be harmful to your fishes. Normally it takes 6 to 8 weeks for this type of filter to reach maturity after it is installed in a newly set up aquarium. It is unnecessary to wait this long before adding fishes, as long as the tank is not immediately heavily populated.

Although undergravel filters are important, there are two problems to overcome when using them in an aquarium containing Rift Lake cichlids. Firstly, cichlids like to dig down to the filters, uncovering them and thereby making them useless. This problem can be avoided by placing netting material over 1 to 1½ inches of gravel which in turn is over the undergravel filter. This netting should be the same color as the gravel for esthetic purposes and can be held down by rocks. Cover the netting with an additional quarter inch of gravel. The second problem to overcome when using an undergravel filter involves pH. Undergravel filters tend to lower the pH in aquaria. This effect can be minimized by care in not overfeeding your fishes and by regular water changes. Overfeeding could either clog your undergravel filter or create an over-active system. Ten to fifteen percent of the water should be changed weekly (semi-weekly for heavily populated tanks) by siphoning loose debris off the gravel while stirring the gravel surface with your siphon hose. Remember to maintain your salinity concentration by adding the appropriate portions of special salts/seawater to your replacement fresh water.

AERATION

In addition to the filtration system in your aquarium, it is a good idea to have an air pump with an airstone to help circulate the aquarium water. Although aeration is not always essential, Rift Lake cichlids are tropical fishes whose water temperature should be 24-27 °C (75-80 °F). Warm water cannot hold dissolved oxygen as well as cold water, so good water circulation is a necessity.

HEATERS AND TEMPERATURE

As was just mentioned, your Rift Lake cichlid aquarium should be 24-27 °C (75-80 °F). For most climates, this would necessitate the use of an aquarium heater. There are many types of heaters, but I would recommend a thermostatically controlled heater that has its heating element submerged below the water line. There is no need for the complete submersion of such a heater, and it is not recommended. However, salts from the aquarium's hard water often cake around equipment at the water's surface. A sealed heater will help prevent shocks. The size (power) of your heater depends on the size of your aquarium and the climate in which you live. As a rule of thumb 3 to 4 watts per gallon of aquarium size (water + gravel + rocks) are necessary for aquaria kept during the winter where the indoor temperature is 60-65 °F; on this basis, a 30-gallon tank should have a 100 watt heater.

LIGHTING

Another important consideration when setting up an aquarium is the type of lighting that you should have. Here there are two routes that you can take; you can inhibit the growth of algae or promote it. Most people consider it too much trouble to scrape the algae off the aquarium glass. Most of the Rift Lake cichlids, however, enjoy scraping the algae off the rocks, as that is their natural source of nutrition when in the lake. The shallow rocks in Lakes Malawi and Tanganyika are covered with a thick mat of algae called by the German name *Aufwuchs*. Most of the cichlids within these lakes pick and scrape at this mat for algae, bacteria, insects and crustaceans living within the algal mat.

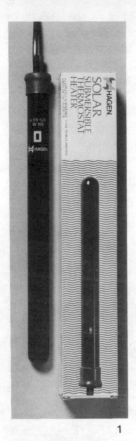

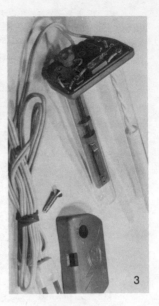

A submersible heater (1) is recommended instead of those which are not water-tight (2-3). Salts from the Rift Lake cichlid aquarium can cake around equipment at the water's surface.

1   2

PLACEMENT

Before setting up your aquarium, it is a good idea to consider its placement. Make sure that your aquarium has a sturdy stand that can take its weight. Many people do not realize how much an aquarium full of rocks, gravel and water actually weighs. Figuring about 10 pounds (4 Kg) per gallon (4 liters), a 30-gallon tank is about 300 pounds when fully equipped and filled with water. It is also a good idea to consider the height at which the aquarium is to be placed. Decide whether you want your tank to be viewed from a sitting position or a standing position, and then place it at eye level for the position chosen. From my experience, it seems that the fish are much more comfortable being viewed from a lateral position than from above.

TESTING YOUR SET-UP

When setting up your aquarium, remember to let your water stand for several days before adding your fishes. Add warm water, as this will help to dissolve the salts in the gravel and let the fluorine and chlorine gases that may be dissolved in the water dissipate. After your water has matured, add a test fish to the tank and wait for two weeks to make sure that there is nothing wrong with your system or the water. Your test fish should be a juvenile or female cichlid; the males are extremely territorial, and later additions will cause problems. If the test fish is a mature male, he will most likely stake out his territory and fight with any new males added to the tank.

ADDING NEW FISHES

If you start with a big tank and decide to add new cichlids at a later date, it is a good idea to completely rearrange the rocks first. Mature males placed into a new tank will challenge each other until they have divided up all the available territory. The most dominant male will command what he considers to be the most preferable area, and the other males will divide up what is left. If a new male is added to the aquarium, he will be abused by the others wherever he may swim. This problem can be rectified by rearranging the rocks within the aquarium and confusing all of the fishes.

DISEASES

African Rift Lake cichlids are susceptible to the diseases that affect most aquarium fishes. This includes mouth fungus and fin and tail rot (slime bacterial infections), ich or white spot (ichthyophthiriasis) and eye fungus (usually a secondary bacterial infection). Most Rift Lake cichlids appear to be parasitized to a certain extent by digenetic trematodes in larval stages. The early signs of these parasites can be seen in the form of small black cysts about the body of the fish, pointing to the area in which the parasite entered the fish's body. Although these parasites affect the coloration of your fishes in a negative manner, they seldom seem to kill their hosts. Furthermore, they do not seem to spread very rapidly from fish to fish in a healthy system.

A relatively common disease among Rift Lake cichlids is one in which the body of the fish seems to fill with water. Termed "Malawi bloat" because of its initial recognition in Lake Malawi cichlids, this disease seems to affect a fish's ability to control its osmotic regulation. The fish literally bloats up with water. Its eyes tend to pop out, its scales stand up, and it loses its balance. Little is yet known about this highly contagious disease, and I would recommend destroying any fish with it.

As this is only a very brief introduction to some of the signs and problems with fish diseases, it is recommended that the hobbyist refer to (or purchase if he does not have) a book on aquarium diseases for guidance.

Feeding is an important consideration for your fish's health. Fish which are fed on inferior products or the wrong foods may survive but will probably not breed. It is best to consider the fish's natural diet. The upper photo (by Glen S. Axelrod) shows *Lamprologus brichardi* in its natural environment in Lake Tanganyika and the lower photo (by H. J. Richter) shows *L. brichardi* in the aquarium.

Feeding Rift Lake Cichlids

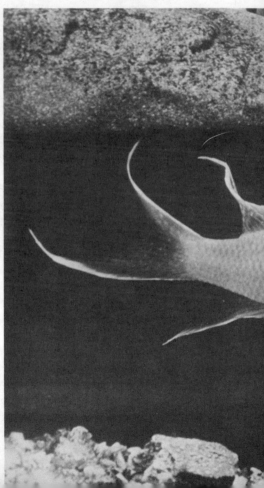

Many scientists stress that the large number of endemic cichlids in Africa's Great Lakes has been caused by adaptation towards different feeding specializations. This includes many of the more specialized cichlid trophic (pertaining to nutrition) modifications of the dentition, oral cavity, pharyngeal bones and digestive physiology. Cichlids' ability to colonize and the diversification of cichlids within the lake environment (in spite of strong competition and predation pressure) is unequalled in its extent by any other vertebrate family. Cichlid trophic specializations include mollusc-crushing, sand-sifting to collect small crustaceans and chironomid larvae, the cropping of higher plants, algae-scraping, insect collecting from algae-covered surfaces, zooplankton feeding, extracting gastropods from shells, scale snatching from other fishes, eating other fishes and extracting eggs and embryos from the mouths of other cichlids.

Because of the extreme morphological differentiation among cichlids, specializations have been developed that exceed the limits ordinarily found within a family of fishes. The leaf-like teeth of *Perissodus* are an excellent example of the extreme specializations found in certain cichlid genera. Another example would be the specializations of the pelvic fins found in the bottom-dwelling *Asprotilapia* and *Xenotilapia*. *Perissodus* are scale eaters and use their leaf-like teeth to pry scales from other fishes. *Asprotilapia* and *Xenotilapia* use their pelvic fins for support so they can "sit" on the lake's bottom. Because of extreme radiative divergence and convergence in different families, some Lake Tanganyikan and Malawian cichlids closely resemble members of other percomorph families such as Girellidae (*Tropheus*), Blenniidae (*Telmatochromis*) and Percidae (*Asprotilapia*).

COMPETITION FOR FOOD

As we have seen, the genetic composition of the cichlids is very "plastic," since modifications are frequent. Selection pressures include the climatic environment, population size, predation and the availability of food. Population size is usually dictated by the other three selection pressures. Most cichlids are territorial and this territoriality also influences population density.

Lake Tanganyika's relatively stable environment has been conducive to speciation. A comparison might be made between the isolated portion of the Malagarasi River and Lake Tanganyika. Few endemic species have evolved within the Malagarasi, and it has been suggested that this is largely because of the river's climatic environment. The alternating periods of drought and flood, to which the Malagarasi has been particularly susceptible, create an unstable environment that is not conducive to specialized forms living in narrow niches. These drastic climatic changes, however, have had less of an effect on the lake, largely because of its enormous depth. During periods of long drought, the water level in the lake dropped, splitting it into two bodies of water. The fish fauna receded with the water level. Thus, more forms of life within the lake were able to survive. The Malagarasi River environment was more favorable to generalized forms. Under stable conditions, foods of particular kinds are in plentiful supply. Specialists that can utilize these particular resources in a more efficient manner than can the generalized forms of life may endure. Harsh climatic conditions, however, may place limitations on the quantity of these various foods. The generalized fishes will have the ability to utilize available food of several kinds. The specialized fishes will probably perish, as they are either restricted to a particular food which is in intermittent short supply or unable to adequately compete with other fishes for the available food. Thus, stable environments are conducive to specialization, while unstable environments will more successfully sustain generalized life forms.

FEEDING YOUR CICHLIDS

The laws of nature and survival of the fittest and best-adapted do not apply in your aquarium. This is because the very nature of the aquarium and the aquarist creates an artificial environment. Your cichlids should be fed enough so that they do not have to compete for food, so their trophic specializations are not so important as they are in the wild. You will find that your fishes will take almost any food, regardless of their habits in the wild. It is up to the aquarist to insure that the fishes in his tank get the proper nutrition.

Feeding should be made often and in small amounts. Most of the current fish foods on the market are acceptable and will be eaten. These include live tubifex worms, freeze-dried tubifex, flake foods, and frozen foods. Larger piscivorous fishes should be given live food if possible. Herbivorous fishes should be given plant matter such as lettuce. Feed no more food than can be eaten in two to three minutes. Remember that cichlids have pharyngeal bones with teeth (teeth in their throat) and that they use this apparatus to macerate their food. Usually they must stuff their mouths with food and then slowly chew it with their pharyngeal bones.

During extended periods of darkness, your cichlids will reduce their activity and appear to "sleep." If you feed your fishes in the morning, give them a chance to "wake up." If possible, feed them four times a day in small amounts. Remember, it is best to feed small amounts often.

In the chapter on "Identification and Selection," I have made some specific recommendations regarding the diet for a few species. Generally, I recommend feeding your fish frozen beef heart two to three times a week. I take a frozen chunk of meat and grate it up before feeding it to the fishes. Once a week I feed some vegetable matter. If possible, keep a second small aquarium in the sun. Place some filamentous algae and macrophytes into the small aquarium. Soon you will have enough algae to feed your fish on a once-a-week basis. For the rest of the feedings, use a flake food. When available, feed with live food as well.

I have found that many of my herbivorous fishes will eat spinach (slightly cooked). I used spinach when I found that I was running low or was out of algae or macrophytes. When one of my non-cichlid aquaria was getting over-populated with plants, I would put the plants into my Rift Lake cichlid tank—they would last a few days before being completely devoured. The fish will also eat insects within plants.

Most of the Rift Lake cichlids are mouthbrooders that lay large eggs and produce large free-swimming fry. Most of the Rift Lake cichlid substrate spawners also produce large fry. A few substrate spawners, however, produce small fry; these fry must be fed on infusoria, as other foods are too large. These small fry are, however, the exception rather than the rule. Most baby Rift Lake cichlids are large enough to eat newly hatched brine shrimp and thrive. Many are even large enough to eat crushed flake food.

Some aquarists do not know the proper way to feed their fry newly hatched brine shrimp. As a result, they kill their fishes. Brine shrimp eggs are widely available and easy to hatch. Detailed instructions come with most packets of eggs. The problems start after hatching has occurred. Not all of the eggs hatch at once, and some never hatch at all. One must be very careful to separate the hatched brine shrimp from the eggs. Take the bubbler out of the brine shrimp hatcher well before collecting the shrimp. Collect the brine shrimp from the middle of the container where there is less chance of contamination by egg shells, etc. (Attract them to the middle by using a light; the nauplii are attracted to light.) If a mistake is made and a substantial number of unhatched and partially developed eggs are fed to the fry, the juveniles may die. The fry will eat the unhatched eggs, which will expand and burst open in the fry's stomach, rupturing it. Care and patience will be necessary to avoid this problem.

1

(1) *Haplochromis fenestratus* grazing on the algae covered rocks in Lake Malawi (photo by Dr. Herbert R. Axelrod). (2) *Spathodus erythrodon* has mouth and body specializations for bottom feeding (photo by Dr. Wolfgang Staeck). (3) *Haplochromis compressiceps* is a piscivore with a characteristically large mouth (photo by Dr. Herbert R. Axelrod). (4) *Petrochromis polyodon* has mouth and tooth specializations for grazing on algae covered rocks (photo by Dr. Herbert R. Axelrod).

2

3 4

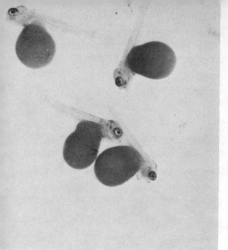

Labeotropheus fuelleborni
after 9 days of incubation (photo by
Dr. D. Terver, Nancy Aquarium).
Melanochromis auratus releasing
fully developed young after the com-
pletion of her mouthbrooding period
(photo by Orsini).

Rift Lake Cichlid Reproduction

The wide diversification of fishes in Lake Tanganyika, which shows more faunal diversification than any other African lake, is probably a result of the lake's greater period of isolation. Geographical isolation is the major reason for the endemic diversity in Lake Tanganyika. Intra-lacustrine speciation has occurred to a much greater extent in cichlids than in all of the other lake fish families (13) combined. This phenomenon is repeated in almost all of the African tropical lakes. Lake Tanganyika contains approximately 140 cichlid species as compared with a total of 67 non-cichlids. Lake Malawi contains approximately 217 cichlid species as compared with a total of 44 non-cichlids. We can only speculate as to the reasons involved. The answers probably lie largely in two areas of cichlid organization: their morphology and their elaborate breeding habits.

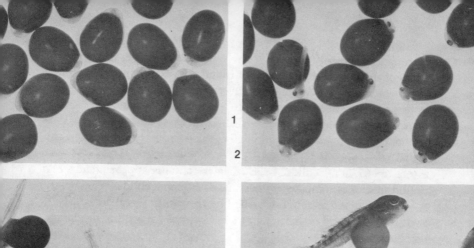

1

2

3

4

5

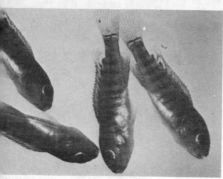

Labeotropheus fuelleborni in the process of development after: (1) 4 days (3½X mag.). (2) 6 days (3½X mag.). (3) 9 days (3½X mag.). (4) 13 days (3¼X mag.). (5) 19 days (3X mag.). Photos by Dr. D. Terver, Nancy Aquarium.

RUNNING WATER

It was essential for the fishes to overcome the problem of lacustrine colonization before adaptive radiation was possible. Spawning was one of the most difficult problems to overcome as it was often closely associated with running water (as cichlids originally came from the rivers). Cichlids evolved the ability to generate their own supply of running water by fanning their fins or, in the case of mouthbrooders, by their opercular movements. This, coupled with increased spawning precision and territoriality, was vital to the success of radiation throughout the lake.

MOUTHBROODERS AND SUBSTRATE SPAWNERS

The pre-spawning activities of the Rift Lake cichlids involve elaborate rituals of courtship, including recognition signs such as color exhibition. The males often build nest areas, and territories are established. These fishes have been traditionally divided into two groups, substrate spawners and mouthbrooders, on the basis of their breeding behavior. Mouthbrooders incubate their fertilized eggs within their oral cavities, and substrate spawners have their fertilized eggs develop on external surfaces. The substrate spawners guard their eggs and developing fry. As some egg guarders lay their eggs in open areas, while others lay them in partially enclosed areas (e.g. stone caves), the terms "open brooders" and "hole brooders" may be applied. While substrate spawners are divided into these two categories, all mouthbrooders are considered to be in the "hole brooder" category. It has been suggested that the substrate "hole brooders" have a closer relationship to the mouthbrooders than to the substrate "open brooders," as the mouthbrooders and substrate "hole brooders" hide their eggs and offspring.

Lamprologus tetracanthus from Lake Tanganyika is an example of a substrate spawner. It is an important example, however, of what can be considered as an early stage in mouthbrooding evolution. Although embryonic development occurs strictly on an external medium (and not in the oral cavity), the parents pick up the young, and often the eggs, to move them to new locations. *L. tetracanthus* displays many of the characteristics found in mouthbrooders. These include a reduction in the egg number to a few eggs of a relatively large size (as compared to many other substrate spawners), an intensified egg color, frequent oral transportation of the young and sexual dichromatism (color differences between sexes) in the adults. Thus *L. tetracanthus* may be considered to be a *brood concealer* and somewhat intermediate between substrate spawners and mouthbrooders.

BREEDING RITUALS

The elaborate breeding rituals of the cichlids present many opportunities for divergence and, hence, for speciation. Slight behavioral changes in the spawning ritual can lead to incompatibility and ethological reproductive isolation (sexual incompatibility due to behavior). It has been suggested that parental care and pair formation of most cichlids would generate micro-communities of inbreeding. These cell communities would tend to impede gene flow and stimulate speciation.

Pseudotropheus sp. preparing a spawning site (photo by Dr. Herbert R. Axelrod). Labeotropheus fuelleborni female nibbling at the genital papilla of the male and collecting sperm to fertilize the eggs in her mouth (photo by Dr. D. Terver, Nancy Aquarium).

Breeding Cichlids In Your Aquarium

One of the most exciting aspects of keeping African Rift Lake cichlids is spawning them. Under the right conditions, many of these fishes will readily spawn in your aquarium. Your success in spawning them hinges, however, on your ability to simulate their natural environment as closely as possible. These conditions were discussed earlier.

I have had the opportunity to successfully breed many African Rift Lake cichlids thanks to a three-year term of study at the J.L.B. Smith Institute of Ichthyology, Rhodes University. Besides the outstanding supervision, the Smith Institute offers excellent facilities including

three constant-temperature rooms, one 2000-liter (500-gallon) display tank, one 1000-liter (250-gallon) experiment tank, four 500-liter (125-gallon) tanks and over forty 100-liter (25-gallon) tanks. All of these tanks had filtration systems, the heat was thermostatically controlled and all of the lights were on time switches. The supervision and equipment, along with several thousand African Rift Lake cichlids, afforded me the opportunity to study the breeding conditions of these fishes in depth.

SETTING UP THE BREEDING AQUARIUM

When setting up aquaria with the intention of breeding Rift Lake cichlids, at least ten gallons of water should be allowed for each fish. The fish can best be bred in 20-gallon tanks accommodating one pair each, or a large tank may be used to breed several pairs. Ignoring esthetics, it is often better for the beginner to place mated pairs in individual tanks. This eliminates the problems of competitive territoriality that the fish will display when they develop the spawning site; it will also eliminate the problem of territorial defense during and after spawning. The breeding tank should not be well planted but must be adequately aerated. Aeration and filtration can best be accomplished in one operation through the use of undergravel filters. Particulate filters can sometimes trap wandering fry. Although it is not always possible, a "mature" tank provides the best chance for success. A mature tank is one that has been in use for six weeks or more. Here the bacteria in the gravel will have reached their exponential plateau and the under-gravel filter will be efficient as a biological filter rather than just a particulate separator. In addition, the tank will have had time to develop its own supply of infusoria, which the newly hatched fry will require. The tank should be supplied with an artificial cave to act as a spawning site. Although not all African Rift Lake cichlids spawn among the rocks, the vast majority, if not all, of the African cichlids available in pet shops do. This cave need not be completely enclosed. Although rocks are preferable, substitutions can be made. A small flowerpot with an opening in the rim is suitable if one cannot find enough rocks. The pot should be half filled with gravel and placed in a rear corner of the tank. A second technique involves placing a gray plastic (non-toxic, of course) pipe, four to six inches in diameter and six inches long, into the tank. Place one end up against the front glass, while the opposite end faces the rear of the tank. Then put an inch of gravel in the pipe. The front of the aquarium should be partially covered with paper so as to obstruct the view into the hoped-for nesting site. When the fish are spawning, one can carefully lift up the paper and peer into the nest.

THE PRE-SPAWNING RITUALS

As was previously mentioned, African Rift Lake cichlids are either substrate spawners or mouthbrooders. Both undergo pre-spawning rituals and post-spawning egg and fry care. Substrate sand spawners often build nests that are several feet in diameter. Naturally, it is very doubtful that this will happen in your aquarium, but it is an illustration of the diligence of these relatively small fishes. Below is a generalized account of the breeding rituals for both mouthbrooders and substrate spawners.

The initial courtship for African Rift Lake cichlids is seemingly ritualistic and very interesting to watch. The male notices the female when her belly is full and protrudes with ripe eggs. The courtship begins with the male chasing the female around the tank. The female is at first elusive and "plays hard to get." This might go on for hours or days. The female will hide from the male among the rocks until she is also ready to breed. Eventually the male will be able to approach the female. The courtship includes fin movements and body twitchings. The two fish stay close together, with the male usually following the female around the tank. Within a short while (if there are more than one pair in the tank), the male will situate himself in a certain area of the tank and defend its "terrain" from intruders. It is important to place the cave nesting areas far enough apart so that the range of territories does not overlap.

SPAWNING

After the initial courtship is over and the territorial range has been established, the male will usually prepare a nesting site. The fishes (usually the males) can often be seen taking mouthloads of gravel from the caves or, in the case of rock substrate spawners, cleaning the rock surfaces so that the adhesive eggs will remain in place. After the nest is prepared, the male will again chase the female throughout the tank. Finally the female will enter the nest with the male in "hot pursuit." Her body trembles during the egg-laying process.

In the case of substrate spawners, such as many *Lamprologus, Tilapia, Boulengerochromis, Julidochromis* and *Telmatochromis,* the female will lay the eggs on the rock surface and the male will pass over them to fertilize. Some substrate spawners will lay their eggs in crater-like nests that are built in the sand. As was previously mentioned, however, these cichlids have not managed to work their way into the aquarium trade. Some *Lamprologus,* such as *L. multifasciatus, L. ocellatus, L. signatus* and *L. wauthioni,* are known to spawn and brood inside mollusc shells.

Mouthbrooders are very abundant in the Rift Lakes (e.g. *Saro-*

therodon, Pseudotropheus, Melanochromis, Tylochromis, Limnotilapia, Petrochromis, Cyathopharynx, Limnochromis, Eretmodus). The females of these genera lay their eggs on the substrate medium and then quickly turn and take them up in their mouths. Several eggs are laid at a time and then retrieved in the mouth. This is done many times over. The male mouthbrooder will often have egg spots on his anal fin. These spots of pigmentation are the shape, size and coloration of the eggs being laid. The female will try to take these into her mouth and, by doing so, will take some of the male's sperm instead and thus fertilize the eggs. Mouthbrooders must lay far fewer eggs than substrate spawners because of the limitation of the size of their mouths as compared to the comparatively limitless substrate space. Curiously enough, however, the mouthbrooders' eggs are also usually much larger than the substrate spawners'. A mouthbrooder will keep the eggs in her mouth until they are completely developed into fry. In the case of Malawian Mbuna, this could be as few as 10 or as many as 60 eggs, 40 being average. *Pseudotropheus zebra* has been observed to lay 17 eggs, *Labeotropheus trewavasae* 10 eggs, *Tropheus* 5 to 10 eggs and *Melanochromis* 25 to 48 eggs. Substrate spawners lay far more eggs; *Tilapia nilotica*, for instance, lays over 1500 eggs. Thus it is obvious why the mouthbrooders are usually the most expensive aquarium fishes to buy—they cannot be bred in mass systems.

In a large cichlid community tank, eggs are often lost to "egg predators." Many of my other cichlids and also my catfish have "learned" that when the spawning ritual takes place, eggs will be laid and they often move in to steal some!

With both mouthbrooders and substrate spawners, the eggs must be fertilized within minutes after they leave the female's oviduct, as the egg case will otherwise harden and become impenetrable to the sperm. The eggs are aerated during their development. In the case of substrate spawners, the eggs are aerated by the female or male as they generate water currents over them with their pectoral fins. In the case of mouthbrooders, which are usually maternal, the eggs are aerated within the mother's mouth by her opercular movements.

Melanochromis auratus are highly territorial and sexually competitive, often defending their territories from fishes four times their size. (1) Two *M. auratus* males. (2) *M. auratus* male in his spawning site prepared by digging out the gravel from under the rock with his mouth. (3) *M. auratus* male who is subdominant to the one in the nesting site. (4) *M. auratus* female brooding with a mouthful of developing eggs. Photos 1 & 3 by Dr. D. Terver, Nancy Aquarium. Photos 2 & 4 by Dr. Herbert R. Axelrod.

1
2
3
4

POST-SPAWNING CARE

Good aeration and filtration are particularly important after the eggs have been laid, especially for the substrate spawners. The eggs are extremely susceptible to fungus, and the developing eggs, with their high metabolism, need a good supply of oxygen. Partial water changes of not more than 25% are recommended weekly for tanks lacking a good biological filter. Otherwise, regular water changes as previously described will be sufficient. These water changes should be continued for several weeks after the young have hatched. Of the total number of eggs, approximately 90% should be fertilized. The unfertilized eggs are not easy to recognize at first, but after a period of time a color difference should become apparent. The fry are placed in a secluded area and are closely watched over for several days.

With both substrate spawners and mouthbrooders, the fry often seek refuge in a parent's mouth. Most Rift Lake cichlids are surprisingly gentle with their own fry and seldom need to be separated from the brood (whereas separation is always essential with many other cichlids). Initially the infusoria that are found in the "mature" tanks are the best food for the young fry of substrate spawners. Baby brine shrimp (or other similar fine live food) should be given to the fry, as it is almost impossible to determine whether the tank has enough infusoria to sustain the fry completely (and it usually has not). The fry will leave the nest area in about three to five days, although they remain close to their parents for an additional one to two weeks. During this time, the parents will herd the stray fry. The fry grow quickly and should double in size within one week.

The family life of the Rift Lake cichlids is especially interesting to watch. There is a cohesive bond between the parents and the fry. After the fry leave the nest, they form a group around one of the parents, usually the mother. If there is a slight disturbance or sensation of danger, the parent will group the fry with jerking movements of its body. The parent and the young will then freeze in position. Often the young will clamber into the mother's mouth to seek refuge. Furthermore, the parents will protect their brood from invading fishes that may enter the family's territorial range.

ARTIFICIAL INCUBATION OF CICHLID EGGS

Mouthbrooders often have difficulty brooding while in captivity; many of them either swallow their eggs or spit them out. This is also a problem with young fishes that have spawned for the first time. In order to help solve the problem of destroyed eggs, especially with rare or expensive species, several types of artificial incubators have been developed. I modified several of these incubator ideas and came up with what I think is an excellent artificial system in which to hatch mouthbrooder cichlid eggs.

To begin with, the fish carrying the eggs must be caught, but only after you are sure that the eggs have been fertilized. This is not always an easy task, especially if there are many rocks in the aquarium. The carrying fish instinctively hides among the rocks and is very defensive. There are several ways to catch the female short of pulling all of the rocks out of the aquarium or spending the entire day chasing it around with a net. If you expect a spawning in your aquarium, place into the tank a flowerpot having a hole in its side, and have the pot rest on a sheet of glass or plastic. Chase the brooding fish into the pot, and then remove the pot plus fish from the tank. Quickly grasp the fish in your palm. (Make sure that your hands are wet before you handle any live fish). Point the head of the fish down and gently rub the buccal cavity (mouth) from below with your thumb. Hold the fish above a net that is touching the surface of the aquarium water but not completely submerged. As you rub the brooding fish under the mouth and it gasps for air, the eggs will spill from its mouth into the net. You can usually look into the mouth to make sure that all of the eggs have been expelled. Return the adult fish to an aquarium where it can recover without being bothered or chased by other fishes. As this process takes a certain developed manual dexterity, I recommend that the uninitiated practice on inexpensive fishes before handling prize specimens.

Submerge half of the net containing the eggs into a small (about 2 gallons) aquarium, making sure that all of the eggs are covered by at least half an inch of water. The chemistry of the water should be the same as that in the Rift Lake cichlid aquarium. The temperature of the water should be 78-82 °F. This tank should not have an undergravel filter or a particulate filter. No filtration system is needed. Fix the net to the aquarium so as to hold it in the pre-described fashion. Run water through an airlift and bubble it over the eggs in the netting so that water gently splashes down on the submerged eggs. Also place an airstone on the bottom of the small aquarium, directly under the net containing the eggs. Gently aerate so that the bubbles hit the bottom of the netting. Mouthbrooder eggs must be constantly aerated and rotated, but in a gentle manner. The combination of slow aeration and splashing water will accomplish this.

Lamprologus compressiceps from Lake Tanganyika is a piscivore which should not be kept with small fishes (photo by W. Hoppe). *Labeotropheus fuelleborni* from Lake Malawi is a relatively peaceful Rift Lake cichlid, but can become more aggressive when breeding (photo by G. Marcuse).

Identification and Selection

Cichlids are bony, perch-like fishes with a body which is usually bilaterally compressed and symmetrical. There are, however, exceptions to this rule. Cichlids comprise an extremely diversified family with incredible morphological variation. They are differentiated from most other perciforms by having one nostril, rather than two nostrils, on each side of the head.

At the moment, all African Rift Lake cichlids of interest to the aquarist come from either Lake Malawi or Lake Tanganyika. During the past ten years, there has been a tremendous revival of interest in these lakes that has not been seen since the beginning of the century. Because of the new techniques of underwater exploration and fish collection that were unavailable to scientists in the past, scores of new species have recently been collected. This has created many taxonomic problems. Additionally, many of these Rift Lake cichlids are very closely related and thus difficult for both layman and scientist to tell apart

without exhaustive examination. As a result, many misidentified and even identified species find their way into pet dealers' tanks. Thus there is plenty of room for error in the chain that finally brings the fishes into your home (collector—exporter—wholesaler—retailer).

The most reliable way to be assured of the purchased fish's name is to first buy it from a reputable dealer in your local area who will back up his sale with well established experience. Valuable information is also available in many African cichlid books on the market which are designed to help you identify your fish (see the bibliography).

In choosing your fishes, there are several points which must be kept in mind: (1) What is available. (2) What you can afford. (3) What you like. (4) How much aquarium experience you have. With the exception of points 3 and 4, you are pretty much fixed to this process of elimination. If you are interested in breeding, try to make an arrangement with your dealer to buy the offspring from you if you should have a successful spawning. If you are planning to buy juveniles or fry, make sure that the ones you are offered are the species that you intend to buy. It is usually very difficult to identify juveniles or fry. Also make sure that you have enough room in your tanks for the fishes when they mature. All in all, you can almost always depend upon your pet dealer for good advice. His business is largely dependent on your success.

For the beginner, I would recommend either *Pseudotropheus zebra* or *Melanochromis auratus*. Both are easily obtainable and very attractive. *Pseudotropheus zebra* has many color varieties and is relatively hardy. *Melanochromis auratus* is very hardy and is easily bred.

The rest of this chapter is devoted to a series of brief introductions to some of the more popular and interesting Rift Lake cichlids. This is by no means a conclusive listing of the fishes available or an in depth review of the species mentioned, but rather some basic information that will be necessary to help you choose and keep your fish. The information covered includes the scientific name of each fish, its common name, a description of its coloration, size, temperament and habitat, comments on sexual dimorphism and breeding, and important notes on its feeding requirements.

Aulonocara nyassae Regan

Aulonocara nyassae, commonly referred to as the African peacock, is a relatively peaceful cichlid from Lake Malawi. Although the males are territorial in nature, the African peacock will usually just "ruffle its feathers" and display its beautiful coloration rather than inflict serious injury on other fishes. This assumes, of course, that they are kept in aquaria of adequate size and are not cramped into close quarters with overlapping territories.

This species can grow up to approximately 15 cm (6 inches) in the aquarium and will begin to show signs of sexual dichromatism (sex color differences) at about 2½ cm (1 inch). The males mature to a shiny dark blue, with brown-red patches of color spread over the anterior body and head. The female keeps her juvenile coloration of brown or grayish brown and has a dozen or so vertical bars along the side of her body. These bars are less prominent in the male. Although the female can develop rusty patches along the forward portion of her body and head, she will always have a relatively bland coloration compared to that of the male.

Aulonocara nyassae is an insect feeder that will take all standard aquarium flakes and basic foods. This fish will thrive especially well, however, on live foods such as tubifex worms, moths, flies or aquatic insects.

The African peacock is a maternal mouthbrooder; it lays few eggs, as they are of a large size. The eggs are laid on a rocky substrate and then picked up by the female, who broods over them for three to four weeks. Upon release from the mother, the free-swimming fry are large enough to eat brine shrimp or finely crushed flake food.

Callochromis pleurospilus (Boulenger)

Callochromis pleurospilus, often called the pastel cichlid, is a peaceful Tanganyikan fish that would make a good member of a Rift Lake community aquarium. As with most cichlids, however, it will eat fishes that it can swallow whole. *C. pleurospilus* is a tannish fish with a bluish silver hue, several irregular patterns of red spots on the upper portion of its body and a red-orange margin on its fins. It seldom grows to more than 9 cm (3½ inches) long. This fish is normally a carnivore, eating invertebrates, other fishes and insects. Nevertheless, it will eat most types of aquarium foods and easily adapts to dry or frozen foods.

Unlike most of the other Rift Lake cichlids that find their way into the aquarium hobby, *C. pleurospilus* does not dwell in a rocky environment; instead, it inhabits sandy shores. Here the pastel cichlid moves about in small schools of largely female fish. The males build crater-like nesting sites in the sandy bottom of the lake. As the females move over the males' territories, the males move up from their respective nests, court a female and then bring her down to the nest to spawn. After the spawning, the female (the species is a maternal mouthbrooder) leaves the nesting area with her fertilized eggs. The male will then look for another female to spawn with.

This species will easily spawn in the aquarium. Only one male is required for several females. Although this is a productive arrangement,

1

2   3

4

5

(1) *Callochromis pleurospilus.*
(2) *C. macrops melanostigma.*
(3) *Aulonocara nyassae.*
(4) *Chalinochromis* sp.
(5) *Chalinochromis brichardi.*
(6) *Cynotilapia afra.*
Photos 1 & 2 by Dr. Herbert R. Axelrod, 3 by Dr. W. Staeck, 4 & 6 by G. Meola, African Fish Imports, 5 by P. Brichard.

6

the females tend to lack the intricate coloration of the males, having a basic silver body color.

Chalinochromis brichardi Poll

Chalinochromis brichardi, often called Brichardi or Brichard's cichlid, is a relatively peaceful Tanganyikan fish that grows up to 15 cm (6 inches) long in nature. Normally found among the shallow rocky outcrops in Lake Tanganyika, *C. brichardi* has only recently found its way into the aquarium hobby. When mature, the fish has a light tan body with black markings on its operculum and preoperculum (gill coverings) and three black bands running obliquely across its head—one on the nape, one between the eyes and continuing through the eyes to the operculum and one above the lip to the eyes. Males are difficult to distinguish from females. Several specimens which have also been referred to as "*C. brichardi*" have a horizontal pattern of black spots, or even sometimes solid stripes, along their flanks. This fish may not be *C. brichardi*, but a new and undescribed species.

Chalinochromis brichardi is an omnivore with unusual structural developments. This fish possesses thick papillose lips which probably serve a sensory function associated with feeding. It is possible that the papillae serve as tactile organs as the fish grazes on algae-covered rocks. In the aquarium, this fish will readily accept prepared dry or frozen foods. Since it is a natural grazer, however, vegetable matter should be included in its diet.

Even though this species is a substrate spawner, it hides its eggs in partially enclosed caves rather than in an open area. It is therefore important to provide adequate rockwork in order to help promote breeding. To date, little study has been done on this species' breeding biology.

Cynotilapia afra Günther

Cynotilapia afra is a rather belligerent fish from waters surrounding Likoma Island in Lake Malawi. It is sometimes incorrectly called the dwarf zebra and should instead be called the dogtooth cichlid. The latter name is derived from the first syllable of its generic name, *Cyno*, which means dog-like (referring to the teeth). The dogtooth cichlid looks very much like *Pseudotropheus zebra*, and the two fishes can easily be confused. *C. afra* is smaller than *P. zebra* and has unicuspid teeth, whereas *P. zebra* has an outer row of bicuspid teeth and inner rows of tricuspids. The male *C. afra* has a blue coloration with six to eight dark vertical bars along the body. The fish can change its color shade, usually depending upon its temperament. Females are a lackluster blue-gray color.

The dogtooth cichlid is a zooplankton feeder but will easily adapt to frozen or prepared flake foods. The fish should have some vegetable matter in its diet. Although it is not a natural piscivore, *C. afra* is aggressive and pugnacious and will not adapt well in a community tank. It should be kept with its own kind in large aquaria.

Cynotilapia afra is a maternal mouthbrooder. Like most other Malawian mbuna, the female lays her eggs on an open hard substrate and then collects them in her mouth. Fertilization takes place when the female mouths the male's genital papilla.

Cyphotilapia frontosa (Boulenger)

Cyphotilapia frontosa, commonly referred to as frontosa, is a deepwater Tanganyikan cichlid with a very peaceful nature. This fish has been known to grow to over 30 cm (12 inches), but it is rarely caught at that size and will probably not grow that large in an aquarium. Frontosa is a light blue fish with five bands on its head and body. The males develop a large hump on the head; this hump grows with age. It has become a very desirable aquarium fish because of its beauty and grace. In spite of its large size, it is unlike most other Rift Lake cichlids in that it is exceptionally peaceful and, although it is territorial, will not defend its territory with the same vigor as will many other cichlids. These qualities, along with its scarcity in the aquarium trade, account for the high price that this fish commands. Although *C. frontosa* is widely distributed throughout Lake Tanganyika, it is found at depths that make its capture difficult. Furthermore, even after the fish are caught, many of them die as a result of pressure problems when they are brought to water's surface.

Cyphotilapia frontosa is a generalized carnivore that eats both fishes and invertebrates. However, it will accept most standard aquarium foods. Personal experience has shown that this species does very well on a diet of beef heart supplemented with flake food.

This species is a maternal mouthbrooder. Passive pre-spawning courtship leads to the production of a small number (15 to 25) of large eggs that develop in three to four weeks.

Eretmodus cyanostictus (Boulenger)

Eretmodus cyanostictus, commonly called the striped goby cichlid, is a Tanganyikan fish that looks and acts more like a goby (Gobiidae) than a cichlid. If kept by itself, it can be maintained in small aquaria. Growing up to 7½ cm (3 inches), *E. cyanostictus* is an awkward-looking fish that "hops" and "jumps" along the rocks on the bottom of shallow shorelines. This fish has a rather bland olive gray coloration which is

1

2   3

4

5

(1) *Cyprichromis leptosoma.* (2-3) *Eretmodus cyanostictus.* (4) *Haplochromis annectens.* (5) *Eretmodus cyanostictus* in its typical habitat in Lake Tanganyika. (6) *Cyphotilapia frontosa.* Photos 1 & 5 by Glen S. Axelrod, 3 by Wardley Products Co., 4 & 6 by Dr. Herbert R. Axelrod.

6

more than made up for by its interesting non-cichlid-like behavior. Best adapted for bottom-dwelling, this species is a poor swimmer and seems to be weighted down. This configuration helps to sustain the fish in the often turbulent shore waters of the Lake. Its pectoral fins are often used for "walking" along the substrate while it drags its pelvic fins and uses them for support. The fish has a sub-terminal mouth (forward and underslung) and therefore must grasp its food from above. This species generally ignores rock-dwelling or midwater-swimming cichlids but will often fight for territory among its own kind. Sexing is difficult.

Eretmodus cyanostictus will accept most prepared and live aquarium foods. The fish will often come to the top during feeding unless there are other aggressive fishes in the tank to inhibit it. When feeding it, remember to consider this fish's small mouth; if it does not come to the top of the tank, make sure that some food reaches the bottom.

This species is a maternal mouthbrooder and is very temperamental. If the female is startled or upset during her brooding period, she may expel or eat some or all of her eggs. The brooding period is 4½ to 5½ weeks for 15 to 25 eggs.

Haplochromis annectens Regan

Haplochromis annectens, commonly called annectens, is an aggressive Malawian cichlid that should not be kept with other fishes that are either small or easily dominated. The males of this species are blue with a mid-body horizontal dark band from the operculum to the base of the caudal. The fish also has faint vertical barring on its flanks. The scales seem to reflect all of the colors of the rainbow, creating a myriad of reflections that give this fish its true beauty. It grows to about 19 cm (7½ inches) but, because of its nature, requires a large aquarium (no smaller than 20 gallons for one fish). The female has a more silvery body.

This fish is a generalized omnivore and will take almost all aquarium foods without a problem. Nevertheless, it is a good idea to provide live food on occasion. This live food should be suitable to the size of *H. annectens* and could include tubifex worms and food fishes.

Haplochromis annectens is a maternal mouthbrooder that will build a nest in the gravel of the aquarium. It lays 50 to 90 large eggs which have an incubation period of between 2½ and 3½ weeks.

Haplochromis compressiceps (Boulenger)

Haplochromis compressiceps, commonly called the Malawi eye-biter, is a predator from Lake Malawi with the unusual reputation of biting out the eyes of its victims. This piscivore will eat small fishes that it can completely engulf. Its name is derived from its compressed morphology, and this fish's appearance also gives it an ominous look that it does not completely deserve. Its reputation as an eye-biter has been largely unsubstantiated. Being a predator, it is possible that this fish may dislodge the eye of its prey while in the course of eating it. An entrapped fish, caught head first, might successfully escape the grasp of its predator but lose an eye in the process. Whatever the story, I find this legend hard to believe. Keeping several large specimens of this species (in a large tank), I have found them to be exceptionally peaceful toward other cichlids, even those only half its size.

H. compressiceps grows to over 30 cm (12 inches) and has an over-all silvery coloration and two dark bands along its flank and back. When mature and in breeding color, males gain a bluish hue and, depending upon the angle of view, show red or green iridescence. It is otherwise difficult to distinguish the sexes. Its body shape and coloration help this fish to camouflage itself among the *vallisneria* beds in which it often hunts. Suspended head down or at an oblique angle, *H. compressiceps* gently moves among the vegetation until it is close enough to strike at its prey.

This species will accept many standard aquarium foods but should on occasion be given live food fishes. This will also help to reinforce its natural behavior. The Malawi eye-biter is a maternal mouthbrooder that spawns like other Rift Lake *Haplochromis* species. Brooding lasts about 3 weeks.

Haplochromis euchilus Trewavas

Haplochromis euchilus, commonly called euchilus or big-lips, is a usually peaceful Malawian cichlid that is less territorial (except when breeding) than most other Rift Lake cichlids of its size. Growing up to 33 cm (13 inches) in the wild, this fish lives in the rocky littoral zone (shallow enough to allow photosynthetic flora to exist) of the lake.

Here *Haplochromis euchilus* grazes on the heavy algae carpeting the rocks, feeding largely upon the insects that in turn live off the algae. Juvenile specimens lack the large lips of the adults. Adults are of a blue-silver color and have two dark bands running along their flanks and back. Sexing live specimens is usually difficult.

In the aquarium, this species will accept freeze-dried or frozen foods, although it usually prefers and thrives on live food. It is often difficult to get this fish to accept even the best flake foods.

Haplochromis euchilus is supposedly an extremely difficult fish to spawn in the aquarium. It is a maternal mouthbrooder that has been known to lay up to 90 eggs which have an incubation period of three to

1

2

(1) *Haplochromis euchilus.*
(2-3) *Haplochromis com-
pressiceps.* (4 & 6)
Haplochromis linni. (5 & 7)
Haplochromis horei.
Photos 1, 2 & 6 by Dr.
Herbert R. Axelrod, 3 by K.
Lucas, Steinhart Aquarium,
4, 5 & 7 by Glen S. Axelrod.

3

4

5

6

7

four weeks. I lost the occasion to spawn a pair of these fish when the female died of the Malawi bloat disease. After six months, the male matured and was trying to court and spawn with other males (mbuna species) in my Rift Lake community aquarium. This homosexual behavior is not completely unusual and has been noted among other species lacking mates.

Haplochromis horei (Günther)

Haplochromis horei, referred to as the spotted or spotheaded haplochromis, is an aggressive fish that will not hesitate to eat smaller fishes. Although it is a generalized feeder, this fish will waste no time in seeking out and devouring other fishes that it can wholly engulf. *H. horei* attains a length of over 18 cm (7 inches) and will accept most types of aquarium food. Ideally, a mixed diet of plant and animal matter should be given, as the fish is naturally an omnivore. This diet can include frozen beef heart, spinach, flake foods and live foods such as brine shrimp and tubifex worms.

H. horei lives in the shallow waters around the coastline of Lake Tanganyika. Its range extends into the river mouths and estuaries that flow into the lake. *H. horei* can be found at depths of less than 3 meters (10 feet) and does not venture out into the deeper waters of the lake. Although little is known about its spawning, it has been reported that craters or bowl-shaped depressions are dug into the sand or silt on the lake's bottom to act as nests. It is assumed that, after a pre-spawning courtship, the eggs are laid in the depression and then picked up into the mouth of the female. As this fish is a maternal mouthbrooder, the female will carry the eggs until development. After maturity, and especially during spawning times, males develop much more radiant coloration than the females do, including the presence of bright red spots that cover his entire body.

Haplochromis linni Burgess and Axelrod

Haplochromis linni, referred to as the elephantnose cichlid, elephantnose polystigma or Linn's haplochromis, is a relatively passive cichlid that will fit in well with a community tank of other large African Rift Lake cichlids. *H. linni* is, however, a piscivore that specializes in eating small juveniles or fry. I have seen it in both Lake Malawi and in my own aquarium alertly perched over a rock with its trunk-like snout hanging over the edge. When the prey comes within striking distance, *H. linni* will literally suck it into its mouth. This fish has been reported to have reached lengths of 35 cm (14 inches). Thus it would be wise to make sure that there are no fishes in your aquarium that are small enough for your elephantnose cichlid to engulf whole.

H. linni's spotted-bloched body closely resembles those of *H. polystigma* and *H. livingstonii* in color and color pattern. As in the other two fishes, *H. linni* becomes much lighter when in aggressive or breeding coloration. Spawning follows the same pattern as with the other two fishes. The female is reported to lay as many as 200 eggs, which incubate in about three weeks. Because of its scarcity, little is known about the breeding habits of this fish. Sexing is difficult.

The elephantnose cichlid will eat a large variety of fish foods. It is recommended that beef heart and live food fishes be fed as well as flake foods.

Haplochromis livingstonii (Günther)

Haplochromis livingstonii, commonly referred to as livingstoni, is a very aggressive fish that should not be kept with smaller fishes or even many of its own kind. Often confused with its close relative, *H. polystigma, H. livingstonii* has large tan-brown blotches of coloration over a white body background. The fish attains a length of up to 30 cm (12 inches) in length in Lake Malawi but will usually not reach this size while in the aquarium. Males grow larger than the females and, when mature, exhibit a blue hue about their head.

Haplochromis livingstonii males are extremely aggressive and very territorial. The fish will fight with its own kind and its close relatives *(H. polystigma* and *H. linni)* more than it will with other fishes in the tank. Therefore it is advisable to provide it with a fairly large aquarium. Extensive rockwork is needed so that the sub-dominant males can escape.

Although *H. livingstonii* will take a wide variety of foods, live food is preferred whenever possible. If live food is unavailable, beef heart supplemented with flake food is acceptable. This fish is a predator by nature and will eat any small fishes that it can catch. In open water *H. livingstonii* will actively chase its prey. Usually, however, the smaller fishes in the Rift Lakes will seek shelter among the rocks. One of the hunting techniques of *H. livingstonii* is to lie in wait amongst these rocks and engulf unsuspecting fry that wander within striking distance. Livingstoni will often be seen lying on its belly or side waiting for its prey.

Spawning is easily accomplished and will often readily occur in a Rift Lake community tank. The male will actively defend his territory during the spawning. This fish is a maternal mouthbrooder. The incubation period lasts approximately three weeks, and broods number up to 200.

(1) *Julidochromis ornatus.*
(2) *Iodotropheus sprengerae.* (3) *Haplochromis livingstonii* (4-5) *Haplochromis moorii.* Photo 1 by H. J. Richter, 2 by S. Frank, 3 by Dr. Herbert R. Axelrod, 4 by G. Schubert, 5 by G. Marcuse.

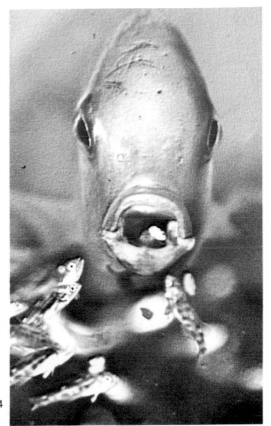

4

5

Haplochromis moorii (Boulenger)

Haplochromis moorii, commonly called the blue lumphead, is a relatively peaceful cichlid from Lake Malawi. It has an over-all (body, head and fins) blue coloration, and black markings are found on certain forms depending upon their location of capture. The blue lumphead is an omnivore, feeding on invertebrates, macrophytes and algae. It grows up to 23 cm (9 inches) and is very difficult to sex. The males usually have a more intense coloration than the females, are larger and have a larger lump on their head. Because of its passiveness, beautiful cobalt-blue coloration and scarcity, *H. moorii* commands a high price.

Haplochromis moorii accepts most aquarium foods, especially frozen or live foods. Vegetable matter should be included in the diet.

This fish is a maternal mouthbrooder with an incubation period that lasts three to four weeks. As with most other mouthbrooders, the fry are free-swimming upon release from the mother's mouth and are large enough to eat newly hatched brine shrimp.

Iodotropheus sprengerae Oliver and Loiselle

Iodotropheus sprengerae, commonly called the rusty cichlid, is a belligerent and territorial fish from Lake Malawi. This fish is especially territorial and destructive towards its own species. This bluish or purplish fish grows to 10 cm (4 inches), with the females being slightly smaller than the males. Although this fish has been in hobbyists' aquaria for over ten years, it was described only in 1972. Previously it was thought to be *Petrotilapia tridentiger* or a *Melanochromis* species.

Although *Iodotropheus sprengerae* will take most aquarium foods, it seems to do best on a diet of meaty foods such as beef heart, live tubifex worms and frozen brine shrimp. The rusty cichlid is usually not a picky eater.

The rusty cichlid is a maternal mouthbrooder that has a brooding time of approximately three weeks. After a successful spawning has taken place, the female should be separated from the male. The female will care for the fry and should be left with them for several weeks after they emerge from her mouth. The mating pair should not be reunited until the female has had a chance to rest, eat and recover from several weeks of fasting while she was brooding.

Julidochromis ornatus Boulenger

Julidochromis ornatus, commonly called julie or ornatus, is a relatively peaceful cichlid from Lake Tanganyika. Nevertheless, this species is territorial, and males must be given enough room so that they do not fight each other. *J. ornatus* is a golden yellow fish with three chocolate-brown stripes in the upper half of its body. It reaches about 7½ cm (3 inches) in length.

This species is omnivorous, eating both algae and insects. In the aquarium, however, *J. ornatus* is a very picky eater. This species seems to prefer live food such as tubifex worms and glass larvae.

Little research has been done on this fish's breeding habits. It is a substrate spawner that lays many eggs in comparison to the small number of large eggs laid by the mouthbrooders in the Rift Lakes.

Lamprologus brichardi Poll

Commonly called the lyretail lamprologus, this Tanganyikan cichlid was formerly called *Lamprologus savoryi elongatus* Trewavas and Poll, but the name was changed, as there was another species preoccupying the name. *L. brichardi* is usually peaceful with other fishes but is often pugnacious with its own kind. The fish is a tan-rust color with long delicate fins that give it an unusual grace. Adults grow to about 10 cm (4 inches), and it is difficult to distinguish the sexes.

This fish is naturally an omnivore and will eat a variety of fish foods. It is especially partial to live foods, however, such as daphnia and brine shrimp. It will also take frozen foods and some flake foods.

Lamprologus brichardi is a substrate spawner that places its eggs in narrow crannies and rock caves. It is often difficult to detect a spawning, as the fish are very secretive. After full development, the fry may not emerge from the rocks for several weeks.

Lamprologus compressiceps Boulenger

Lamprologus compressiceps, commonly called the compressed cichlid or compressiceps (not to be confused with *Haplochromis compressiceps* from Lake Malawi), is a quiet and peaceful Tanganyikan cichlid. Toward its own kind, however, it can be very hostile, and it is recommended that this fish be given plenty of room if several are to be placed together. Also, it will eat small aquarium fishes. There are three known color morphs of this species, the most common one being dark brown with 8 to 12 lighter vertical bars on its sides. A second color variety is deep red, and a third is yellow. It has a flat compressed body and head (hence its name), a very deep body and a large mouth. Adults grow to 11½ cm (4½ inches).

This fish is as fragile as it is graceful. It is very susceptible to disease and is also a very picky eater. It does best on living foods such as tubifex worms, large fully grown brine shrimp and small fishes. In the lake it feeds on small fishes and copepods.

Not a great deal is known about the breeding habits of this species. It

1

2   3

4

5

(1) *Limnochromis auritus.*
(2) *Lamprologus compressiceps.* (3) *L. brichardi.* (4) *Labeotropheus fuelleborni,* red top marmalade cat. (5) *L. fuelleborni,* plain marmalade cat. (6) *L. trewavasae.* Photos 1 & 6 by Dr. Herbert R. Axelrod, 2 by Wardley Products Co., 3 by H. J. Richter, 4 & 5 by G. Meola, African Fish Imports.

6

appears to be a substrate spawner that spawns among the rocks in the littoral region of the lake. It prefers privacy; mating takes place in narrow and enclosed caves. It is extremely difficult to spawn in the aquarium unless the conditions are ideal.

Labeotropheus fuelleborni Ahl

Labeotropheus fuelleborni, commonly called Fuelleborn's cichlid or fuelleborni, is one of the most peaceful cichlids from Lake Malawi. This species is very similar in appearance to *Labeotropheus trewavasae* and *Pseudotropheus zebra*, both also from Lake Malawi. There are several color varieties of this species, the most common variety being blue with 9 to 12 vertical bars on its sides. The species is normally very peaceful. The males often fight, however, when in direct competition for territory or a female.

This fish is an herbivore with very specialized dentition for grazing algae off the substrate. The majority of aquarium foods, including meat, will be accepted, however. Nevertheless, it is important that this fish be fed a good amount of vegetable foods and plant matter.

Labeotropheus fuelleborni is a maternal mouthbrooder with an incubation period of 3 to 4 weeks. Broods range in size from 25 to 50 offspring and are fairly large upon release. Adults grow to over 13 cm (5 inches), with the males being slightly larger and more colorful than the females.

Limnochromis auritus (Boulenger)

Limnochromis auritus, not as yet known by any widespread common name, is a peaceful and shy Tanganyikan cichlid that does not exhibit the usual aggressiveness of most of the other Rift Lake cichlids. It is a beautiful fish with delicate coloration but is very difficult to keep in the aquarium. Its colors vary with its mood from a tan body with a white belly to a beautiful silvery blue body highlighted with golden-orange patches. This fish grows to 14 cm (5½ inches). It is very difficult to distinguish the sexes, even during spawning.

This species in the wild is a small mollusc eater but will readily accept most prepared aquarium foods. Live foods should also be given on occasion. As this is a shy fish, make sure that it gets its fair share of the food and that the other fishes do not eat it all.

Limnochromis auritus is a maternal mouthbrooder but differs greatly from most other Rift Lake cichlids in the number of eggs that it produces and in its post-mating behavior. *L. auritus* can produce up to 400 eggs in a single spawning, far above the number normally produced by a mouthbrooder, especially one of its small size. Additionally, the number of eggs is so large that the female is often unable to carry them all. It has been reported that the male will also pick up some of the eggs if the female cannot handle them all. Although the male is more inclined to eat the eggs than is the female, this behavior is still unusual for Rift Lake cichlids and worthy of note. It is also said that the male will often help to protect and move the brood after the fry are free-swimming.

Melanochromis auratus (Boulenger)

Melanochromis auratus, commonly called auratus or the Malawi golden cichlid, is a very territorial, aggressive fish from Lake Malawi. This species is born with a bright golden yellow color. Juveniles have two black stripes on the upper portion of their sides and another black stripe through the dorsal fin. As the fish get older, bright blue streaks line the black stripes. At maturity, the males change color and become almost the inverse of the female and juvenile coloration. The males are black with golden stripes running along their bodies. This fish grows to 10 cm (4 inches), with the males being slightly larger than the females.

Melanochromis auratus in the wild feeds on algae and the insects and crustaceans that live within the algae. Nevertheless, this species will accept most prepared aquarium foods, including those largely composed of animal matter. Vegetable matter should be included in their diet.

This species is a maternal mouthbrooder with an incubation period of 3 to 4 weeks. The size of its broods ranges from 12 to 40, depending upon the conditions of the aquarium and the age of the fish. Fish carrying their first broods usually have few fry. It should be noted that this species is very aggressive when mating. Males always seem to be looking for a mate and will disrupt fish three times their size to prevent them spawning in their territory. This species was formerly called *Pseudotropheus auratus.*

Pseudotropheus lombardoi Burgess

Pseudotropheus lombardoi, commonly referred to as lilancinius or kennyi, is an aggressive mbuna from Lake Malawi. Being very territorial in nature, this species should be provided with plenty of room and rocks for caves. This is one of the most beautiful fish from Lake Malawi. The males are a deep orange-yellow with faint black bars. The females and juveniles are blue with dark black bars on their sides. At maturity, the males change color in gradual stages that last from 2 to 4 weeks. This species grows to 13 cm (5 inches), with the males being slightly larger than the females.

Lilancinius in nature grazes upon the algal carpet that covers the rocks in the littoral region of the lake. It is an omnivorous fish that eats

1

(1) *Pseudotropheus zebra.* (2) *Melanochromis auratus.* (3) *Pseudotropheus lombardoi* male. (4) *P. lombardoi* female. (5) *P. tropheops.* Photo 1 by Ken Lucas, Steinhart Aquarium, 3 & 4 by Ed Isaacs, Pet Gallery, 5 by K. Paysan.

2

3

4   5

both the algae and the insects and crustaceans that live within the algae. In the aquarium, *P. lombardoi* will accept most types of aquarium foods. For best results in this fish's care, quality frozen foods and vegetable matter should be fed often. Live food is also well received.

This species is not difficult to spawn if adequate conditions are maintained. *Pseudotropheus lombardoi* is a maternal mouthbrooder with an incubation period of 2 to 3 weeks. A usual spawning will consist of between 30 to 40 fry.

Pseudotropheus tropheops Regan

Pseudotropheus tropheops commonly called tropheops, is one of the most peaceful mbunas from Lake Malawi. Over ten color varieties (color morphs) of this species are known. Some of the color varieties are much more attractive than others. The most popular morphs are those that are bright yellow (golden tropheops), light blue, or orange-blotched (OB tropheops).

This species is found in the shallow littoral region of the lake and dwells among the rocks. Besides using the rocks to spawn and live among, this species also eats off them (as do most other mbuna species); it grazes upon the algae-carpeted rocks for its nourishment. Tropheops does well on a wide variety of aquarium foods, including flake food. It is one of the best Rift Lake cichlids for the beginner in this area of the hobby. *P. tropheops* grows to about 10 cm (4 inches), with the males being slightly larger than the females.

Pseudotropheus tropheops is a maternal mouthbrooder with an incubation period of 3 to 4 weeks. Brood sizes of between 30 and 40 fry are normal.

Pseudotropheus zebra (Boulenger)

Pseudotropheus zebra, commonly called the zebra or the Nyasa blue cichlid, is an mbuna from Lake Malawi and is known to have over a dozen color varieties. Its common name came from its most prevalent color form, which was exported from Lake Malawi in 1964. It has a blue body of varying shades with 6 to 8 vertical bars on its sides (BB morph). As additional collecting work was performed around the lake, many new color varieties were found. These forms include: a solid red morph, called the red zebra; a solid blue morph, called the cobalt blue; a solid green morph, called the green zebra (this morph also is often called a cobalt, as the cobalts can vary in color from blue to greenish blue); an orange-blotched morph called the OB; a mottled form called the marmalade cat; a white morph; a yellow morph; a white form covered with black freckles, called the peppered zebra; a solid orange form called the tangerine form; and an albino form.

This species is usually very pugnacious and territorial. It should be given plenty of room in which to move around and establish its territory. It grows to over 10 cm (4 inches). Because there are so many forms, it is not always easy to sex unless one is familiar with all or most of the color varieties.

Pseudotropheus zebra is omnivorous and will eat most aquarium foods, including flake food. It is best to give this fish a balanced diet of both animal and vegetable matter.

This species is a maternal mouthbrooder with an incubation period of between 3 and 4 weeks. Normal brood sizes are between 30 and 40 fry.

Simochromis diagramma Günther

Simochromis diagramma, commonly called diagramma or the diagonal bar *Simochromis*, is an aggressive cichlid from Lake Tanganyika. This is a relatively large fish which grows to a length of 25 cm (10 inches). It is more aggressive with its own kind than with other species and is therefore usually suitable for large Rift Lake cichlid community aquaria. This fish has a gold and silver body with 10 or 11 diagonal black bars on its sides. Females are smaller than the males and have a duller color.

S. diagramma, like all other *Simochromis*, has specialized dentition for grazing algae off the rocks in the lake. In addition to eating the algal growths themselves, *S. diagramma* also eats life forms that live within the algal carpeting. Its food requirements are not critical, and this species will eat most aquarium foods. For best results, however, vegetable matter should be included in this fish's diet.

Simochromis diagramma is a maternal mouthbrooder that spawns among the rocks in the littoral region of the lake. The eggs have an incubation period of 3 to 4 weeks. Depending upon the age of the fish and its size, this species produces between 40 and 80 eggs at a time.

Telmatochromis caninus Poll

Telmatochromis caninus, commonly called caninus, is a fairly quiet cichlid from Lake Tanganyika. This species lives among the rocks and usually will not be much of a bother to other species. This fish can fight among its own kind, however, especially when spawning. *T. caninus* has a tannish-gold body with rust colored markings that change intensity depending upon the fish's mood.

This species can often be a picky eater. It prefers live or frozen foods along with some vegetable matter. Flake food will usually, but not always, be taken.

1

2

(1-2) *Telmatochromis caninus.*
(3) *Simochromis diagramma.*
(4) *Tropheus duboisi.* (5) *T. moorii.* (6) *T. polli.* Photos 1 & 2 by H. J. Richter, 3, 4 & 6 by Glen S. Axelrod, 5 by Dr. Herbert R. Axelrod.

3

4 5

6

Telmatochromis caninus prefers to stay near the bottom of the aquarium and dwell among the rocks. It is a substrate spawner that will usually lay 50 to 100 large eggs under rock formations. This fish will guard its eggs and protect the fry after they hatch.

Tropheus moorii Boulenger

Tropheus moorii, commonly called moorii, is a relatively peaceful Tanganyikan cichlid that has become very popular. This fish inhabits the rocks in the littoral region of the lake. Over 30 color varieties of this species have been found around the lake, possibly more color morphs than has been found in any other Rift Lake cichlid. These color varieties include an olive form, yellow-bellied form, brown form, green form, red-striped form, orange-striped form, rainbow form, black form, solid orange form, yellow form, blue-black form, red and yellow form and tangerine form. This species grows up to 10 cm (4 inches). It is peaceful with other fishes but can be rough with its own kind if crowded into too small an area or when breeding. Sexing is difficult unless one knows the color morphs that one is working with. Males are usually larger than the females and more colorful.

Tropheus moorii is an omnivore that eats both algae and insects from the rocks. It has specialized dentition very similar to that of the *Simochromis* species. This fish will take most aquarium foods without question and should be given vegetable matter as a supplement to its diet.

This species is a maternal mouthbrooder that lays few (6 to 12) comparatively large eggs. These eggs are about 3/16 of an inch in diameter. Incubation periods range from 4 to 5 weeks. It is important not to disturb this species when brooding, as it is temperamental and will eat or spit out its eggs. This nature, along with its long incubation period and small egg number, helps account for this fish's high price.

Tropheus duboisi juveniles have white spots on their bodies which disappear when they mature. Photo by W. Hoppe.

It is impossible to discuss all of the 700 Rift Lake cichlids within the covers of this small book. Likewise, it is impossible to go into great detail about their biology and life. This section included discussions of more than 20 Rift Lake cichlids (of 16 different genera) that have found their way into the aquarium hobby. Most of the previous are very popular fishes. Other Rift Lake cichlids that can be obtained for your aquarium include *Genyochromis mento* (L. Malawi), *Haplochromis labrosus* (L.M.), *H. polystigma* (L.M.), *H. rostratus* (L.M.), *Julidochromis dickfeldi* (L. Tanganyika), *J. regani* (L.T.), *J. transcriptus* (L.T.), *Labeotropheus trewavasae* (L.M.), *Labidochromis vellicans* (L.M.), *Lamprologus elongatus* (L.T.), *L. furcifer* (L.T.), *L. tretocephalus* (L.T.), *Melanochromis johanni* (L.M.), *Pseudotropheus fuscus* (L.M.), *Pseudotropheus elongatus* (L.M.), *P. livingstonii* (L.M.), *P. microstoma* (L.M.), *Spathodus erythrodon* (L.T.), *Telmatochromis bifrenatus* (L.T.), *Trematocranus jacobfreibergi* (name in doubt, L.M.), *Tropheus duboisi* (L.T.) and *T. polli* (L.T.). Additional reading, which would include information on most of these species, is listed at the end of this book.

Additional Reading

Below are some additional books from T.F.H. Publications, Inc. which contain information about the African Rift Lakes and their fishes.

Axelrod, H.R. 1978. *Beginning with Mbunas.*

Axelrod, H.R. & Burgess, W.E. 1979. *African Cichlids of Lakes Malawi and Tanganyika.* 8th ed.

Axelrod, H.R., Emmens, C.W., Sculthorpe, D., Vorderwinkler, W., Pronek, N., & Burgess, W. 1977. *Exotic Tropical Fishes.* 26th ed.

Brichard, P. 1978. *Fishes of Lake Tanganyika.*

Fryer, G. & Iles, T.D. 1972. *Cichlid Fishes of the Great Lakes of Africa.*

Goldstein, R.J. 1970. *Cichlids.*

Goldstein, R.J. 1973. *Cichlids of the World.*

Goldstein, R.J. 1971. *Introduction to the Cichlids.*

Jackson, P.B.N. & Ribbinck, A.J. 1975. *Mbuna: Rock Dwelling Cichlids of Lake Malawi, Africa.*

Vierke, J. 1979. *Dwarf Cichlids.*